The Crypto Directory

AUDE

1

An Aude™ Publication

First Print Edition 2021.

Hardcover ISBN 978-1-957470-01-6
Paperback ISBN 978-1-957470-00-9
eISBN 978-1-957470-02-3

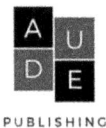

PUBLISHING

To those who create.

Contents

Preface

This book is a guide and a tool for those acquainting themselves with cryptocurrency. New technologies often prelude massive opportunity, yet only a tiny sliver of people take full advantage of these opportunities. These people are characterized by early exposure followed up with a willingness to learn and take on risk. Just consider the early internet builders of the 21st century, the manufacturing tycoons of the Asian Miracle, the colonizers of the United States, and the early industrializers of Britain.

Few fields meet these precedents as much as cryptocurrency. Cryptocurrencies have created millions or billions in wealth for a sizable group of early adopters, yet most other people are not familiar with the field nor technology. Many are beginning to see that cryptocurrency, decentralized applications, and blockchains are enabling the largest financial revolution since the internet, and these technologies are changing the way money and people operate on a day-to-day basis. This book and the knowledge within are representative of a future surely headed our way.

The conceptual barrier of entry to cryptocurrency, in essence, is the driving purpose behind the creation of this work. Learning is fast-tracked through concept explanations, section summaries, visuals, and more. That said, not all need be known; selecting *what* you learn is almost as important as *if* you learn. To meet this expectation, the book is split into two parts: one, the guide, and two, the dictionary. The guide (ending on page 182) can be read as an avenue towards a full base of understanding. Following the guide, the dictionary and resource sections function as reference tools and a springboard for independent learning. Once completed, you should be able to confidently apply yourself in the crypto world. That is the goal, and that is this book.

The guide portion consists of the following sections:

Essential Terms & Concepts

General Knowledge Terms

Blockchain Terms

Trading Terms & Concepts

Informal Language

Acronyms

Visual Catalog

Key Players

History & Timeline

Legality

These sections aim to provide a complete introduction to cryptocurrency. Note that a large number of definitions are provided only in the dictionary for use as a tool of reference, while the guide is less technically inclined.

I'll sign off here. Onwards!

Essentials

The following section forms the core and essential layer of knowledge necessary to operate in the cryptocurrency ecosystem.

This section is separated into two sub-sections: concepts and terms. The following seven concepts outline some of the essential ideas and applications of the wider space and industry of cryptocurrency:

> #1: Decentralization
> #2: Blockchain
> #3: Cryptocurrency
> #4: Cryptocurrency Mining
> #5: Bitcoin
> #6: Ethereum
> #7: NFTs

The terms section contains around fifty words. Terms can be read in any order and are indexed and searchable in the full dictionary at the end of the book.

Decentralization

Cryptocurrency essentially aims to solve the issue of trust. Centralized networks base themselves around some kind of central structure. To the payment sector, this is Paypal and Stripe. To the financial sector, this is a small number of large banks. In the data sector, this is Google, Facebook, and a small number of other large search engines and social media applications.

In each of these systems, control is placed in the hands of a small subset of people or technology, and access is limited. This inherently leads to a multitude of issues: subjectivity, single points of failure, imbalanced decision-making and awareness, and lack of individual incentive to make change from the bottom-up. Additionally, centralized systems often lack, at the very least on a relative scale, transparency, security, and trackability.

Decentralized networks, which operate through distributed networks of nodes (devices on the network), offer all those things: enhanced security with no single point of failure, improved efficiency through automation and removal of various centralized bottlenecks, complete transparency and trackability through a public system of record, decision-making open to nearly all parties in the network, and so on.

This isn't to say the question of centralized versus decentralized systems and organizations has a fully correct answer: both have pros and cons. For example, it is fair to say that centralized control in times of war is a much better system; one adept general is much more efficient than an upper command of a dozen. Yet, by this same mark, the decision on whether to go to war in the first place is much better made in a decentralized manner and with the combined output of many different opinions.

In the past, centralized systems have dominated, and human networks generally trend towards the centralized: many farmers to one city, many political parties to two, movie theaters to a few streaming services, and so on. The specific types of decentralized and distributed networks and the benefits enacted by such networks have only become viable in the past decade or so by the discovery and implementation of new technology, and for that reason decentralized systems are being implemented to massive effect all over the world and in all fields of the world.

The specific type of technology that has promoted this growth is blockchain. Blockchain networks are essentially a new type of database and a new means of transacting. They do this through decentralized networks of computers that connect to shared systems—the Bitcoin network, for example, operates upon many thousands of individual nodes.

Data is encrypted through a hashing process (which essentially scrambles text) and is validated by nodes on the network. In this way, even if individual nodes are corrupted, attackers cannot breach the system since many other uncorrupted nodes simply check and invalidate the information that the corrupted parties are delivering to the system.

When information is transacted on a blockchain (e.g., when party A sends money to party B), information asymmetry is leveraged through the use of a private and public key. Even though the attacker knows the system and the means by which the data is transacted (data meaning the encrypted transmission), the attacker cannot break the system because party A and party B, through code on the blockchain, leverage their private keys to successfully transmit data. All of this is explained in further detail throughout the book.

Blockchain networks require no trust whatsoever, since the code of the system is unalterable, and records are completely public and viewable through public ledgers. No middlemen are needed, overhead, transaction costs, and errors are reduced, and systems are more efficient than centralized alternatives. These benefits have led to the widespread adoption of blockchain across supply chain logistics, data management and verification, digital identity, decentralized web applications, and many more fields.

Blockchain

All cryptocurrencies operate through blockchain technology. Blockchain, in its most basic form, can be thought of as a type of network that stores data in literal chains of blocks. Blockchains are faster and more secure than most centralized alternatives. Here is exactly how blocks and chains come into play:

♦ Each "block" stores digital information, such as the time, date, amount, etc.

♦ The block stores the identity of participants in a transaction by using "digital keys," which are strings of numbers and letters received when opening a wallet. Wallets provide access to crypto assets, just like a bank account.

♦ However, blocks cannot operate on their own. Blocks need verification from other computers, aka "nodes" in the network.

♦ The other nodes will validate the information of one block. Once they validate the data, and if everything looks good, the block and the associated data will be stored on the public ledger.

♦ The public ledger is a database that records every single approved transaction ever made on the network. Most cryptocurrencies, including Bitcoin, have their own public ledger.

♦ Each block in the ledger is linked to the block that came before it and the block that came after it. Hence, the links the blocks form create a chain-like pattern, and a blockchain is formed.

Summary: The **block** represents digital information, and the **chain** represents how that data is stored in the database. Consider viewing the cryptocurrency timeline further along in the book for a look at how blockchain has advanced over the decades.

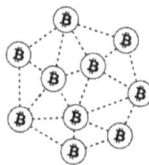

Cryptocurrency

Cryptocurrency is a blanket term. Just like an app store, each app serves a different purpose, and the term "app" merely serves to describe their nature of being a cryptocurrency is a digital system that operates as a medium of exchange by utilizing coins or tokens. Ownership records are stored in a ledger, which is a database that leverages cryptography to enable security and control supply.

Cryptocurrencies can work through a variety of different models: Proof-of-Work (PoW) leverages a defined amount of computational effort to verify transactions, while in Proof-of-Stake (PoS), token owners stake their tokens as collateral. Cryptocurrencies are most notable relative to fiat currencies for being decentralized, and continue to see immense growth in global adoption. Consult the cryptocurrency timeline further on in the book for information regarding the emergence of cryptocurrencies. Shortly upcoming are breakdowns of the main function of PoW networks (mining), as well as the two most popular cryptocurrencies, Bitcoin and Ethereum.

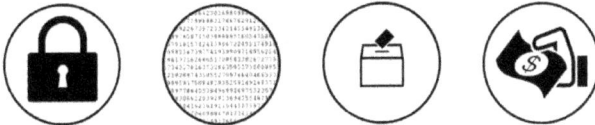

Cryptocurrency Mining

Mining a cryptocurrency is the process of a computer, graphics card, or Application Specific Integrated Circuit (ASIC) trying to discover the mathematical answer to cryptocurrency blocks. Each miner competes in a race to find the answer against all of the other miners on the network. Once the answer is found, that miner is allowed to generate the block and submit it to the network for verification and validation, which is done by the nodes. All Proof-of-Work based cryptocurrencies use some form of mining with a variety of mining algorithms. Bitcoin uses the SHA-256 algorithm, which stands for Secure Hashing Algorithm and outputs 256 bits. (Each bit being a 1 or a 0.) Ethereum uses the EthHash mining algorithm, which is primarily mined used GPUs. The mining market is worth several billion dollars annually, and continues to see growth as crypto mining companies like Argo Blockchain and HIVE go public at 9-figure market caps.

Bitcoin

Bitcoin was the first widely used cryptocurrency: it is an open-source, peer-to-peer computer network, a collection of protocols, and a digital gold. In physical form, Bitcoin operates through 13,000 computers loaded onto the network. In purpose, Bitcoin is a global means of easy and secure transaction, a democratizing force, and a means of transparent and anonymous finance.

Bitcoin was started by Satoshi Nakamoto, who published the Bitcoin white paper in 2008 and has since remained completely anonymous. Regardless of his identity, the creator of Bitcoin currently holds a fortune worth more than $70 billion (equivalent to 1.1 million bitcoins) and if Bitcoin goes up another few hundred percent, this mysterious figure will become the richest individual in the world. The following factors signify some of the benefits of Bitcoin and enable the high degree of security that Bitcoin maintains:

Bitcoin is public. Bitcoin, like many cryptocurrencies, has a public ledger that records all transactions. Since no private information must be provided to own and trade Bitcoin and all transaction information is public on the blockchain, intruders have nothing to hack into or steal; the only alternative to hacking into and profiting off the Bitcoin network (excluding human points of failure, such as in exchange attacks and lost passwords) is a 51% attack, which, at the scale of Bitcoin, is practically impossible. Being public also ties into Bitcoin being permissionless; no one controls it, and therefore no subjective or singular viewpoint can affect the entire network without the consent of everyone else on the network.

Bitcoin is decentralized. Bitcoin currently operates through 13,000 nodes, all of which collectively serve to validate transactions. Since the entire network validates transactions, there is no way of altering or controlling transactions unless, again, 51% of the network is controlled. Such an attack, as mentioned, is practically impossible; at the current price of bitcoin, an attacker would need to spend many millions of dollars a day and control a volume of computational resources that simply isn't available. Hence, the decentralized nature of data validation makes Bitcoin extremely secure.

Bitcoin is irreversible. Once transactions in the network are confirmed, the confirmation is permanent since each block (a block being a batch of new transactions) is connected to blocks on either side of it, hence forming an interconnected chain. Once written, blocks cannot be modified. These two factors, in combination, prevent data alteration and ensure greater security.

Bitcoin uses the hashing process. A hash is a function that converts one value into another; a hash in the crypto world converts an input of letters and numbers (a string) into an encrypted output of a fixed size. Hashes help with encryption because solving each hash requires working backwards to solve an extremely complex mathematical problem. Hence, the ability to solve these equations is purely based on computational power. Hashing algorithms can compare hashed values as opposed to comparing data in its original form, which is much more laborious, and hashing functions serve as one of the most breach-proof means of data transmission.

Using the technologies described, Bitcoin aims to do the following:

- Give users the ability to transact value over the internet in a secure fashion without relying on a central institution, instead relying on cryptographic proof.
- Eliminate the need of intermediaries and decrease friction in supply chain, banking, real estate, law, and other fields.
- Eliminate the dangers faced by the inflationary, centralized, and alterable environment of fiat currencies.
- Enable secure control over personal assets without relying on third-party institutions.
- Enable blockchain solutions in medical, logistical, voting, and finance fields, in addition to wherever else such solutions may apply.

Bitcoin is the most-known cryptocurrency, and rightfully so: it was the first successful crypto, dwarfs all others in size, and incurred rapid mainstream adoption. Still, it represents just a sliver of cryptocurrencies, and many more digital currencies were created to improve upon Bitcoin or to solve other problems. Ethereum, the second-largest cryptocurrency, is one such case.

Ethereum

Ethereum is an open source, blockchain-based technology platform, consisting of a network of over 2,900 computers working together to store and maintain a transaction ledger at a rate of up to 30 transactions per second. The technology behind Ethereum enables privacy, anonymity, and transparency in the world of virtual finance. More than this, the Ethereum platform enables hundreds of thousands of independent cryptocurrency ecosystems, which run on top of it. Called 'tokens', these independent objects can be bought, sold, generated, burned, and transferred using the Ethereum platform. Note that the term "coin" specifically refers to a token run on its own blockchain.

Ethereum also introduced 'smart contracts' which empower a range of applications for both the Ethereum currency and the grander token ecosystem. The concept of the Ethereum platform was first put forward by a Russian-Canadian programmer named Vitalik Buterin in 2013. Through 2014, Buterin worked with a group of individuals to form what is today known as the Ethereum Foundation. Development was crowdfunded, with investors purchasing Ether (ETH) with bitcoin (BTC) in order to get the project off the ground. Buterin's involvement with cryptocurrency began in 2011, when he started writing for *Bitcoin Weekly*, and later for *Bitcoin Magazine,* of which he was a co-founder. His involvement in the Bitcoin community helped to ground him firmly in the concept of cryptocurrency and inspire him toward what he viewed as a once-in-a-generation opportunity.

While Ethereum is a public, decentralized, and transparent organization, Buterin's involvement with the project has given him something of a figurehead status, and his opinions and activities hold some sway over the financial performance of the network and underlying token. Since Ethereum's inception, the Ethereum Foundation, working with Buterin, has driven the technical development of the network. The Foundation's process of suggesting capabilities to the network has brought forward now-ubiquitous features such as token sidechains, sharding, NFT capabilities, and more.

Ethereum consists of wallets, a currency (Ether), blocks, nodes, and miners. Each produced block contains a record of transactions made; the blocks are stored on the nodes and created by the miners. On top of this rides a programmable platform, the Ethereum Virtual Machine (EVM), which takes the classic distributed ledger capability of cryptocurrency and transforms it into a massively distributed state machine. This means that programmable computer code can be executed by the thousands of Ethereum nodes active at any one time.

Each block on the chain not only stores transactions in the standard sense, but also the "state" of the blockchain. Most commonly the "smart contract" functionality allows the network to intelligently create and perform transactions based on pre-defined programming. While traditional cryptocurrencies require an endpoint (node or client) to generate transactions, the EVM allows programmed functions to be stored and executed by the network as a whole. In this way, there is consensus regarding the functions being run.

As an example, the following is possible with a smart contract function:

- Create a token with a maximum supply of 1,000,000 tokens.
- Distribute the token evenly to 100,000 wallets.
- On every transaction, send 0.5% of the value to a "burn" address and distribute 0.5% of the transaction to all current holders of the token.

These types of rules are only possible with the distributed computing that Ethereum enables and would not be possible with a classic ledger-type cryptocurrency.

From its inception, Ethereum has functioned around the Proof-of-Work (PoW) mechanism for processing of blocks and rewarding participation in the network. Mining systems compete to produce the next block in the network, which will contain all of the current transactions. The network adjusts difficulty to allow the mining workstations to produce a new block roughly every twelve seconds. If more systems begin mining, the difficulty will increase in order to maintain this desired timeframe. The winning miner will have around 2 ETH credited to their wallet and will also be reimbursed for gas fees incurred by the transactions included in the block.

With the eventual advent of Eth 2.0, the Ethereum network will move to a Proof-of-Stake model, in which expensive and power-hungry mining systems will no longer be needed. Instead, those wishing to participate in the network must stake (deposit) an amount of ETH in order to become a "validator".

Validator nodes will perform the same function as the mining system. The network will randomly select a validator to produce a given block, while the other validator nodes not selected will confirm the block. The chosen validator will receive the reward for that block.

The new PoS model also allows for a technique known as "sharding", which implements multiple chains on the network. Multiple chains can be operated simultaneously to push the transaction-per-second power of the entire network into the hundreds of thousands.

The titans of cryptocurrency.

NFTs

Cryptocurrencies are fungible, which means they are not unique. Each token is the same as the next, and millions or billions of identical coins or tokens can exist. On the other hand, NFTs are non-fungible, meaning that each token is unique and noninteroperable. The difference is the same as fiat currencies and valuable art—both hold value, they just differ in that a piece of valuable art is unique, while a currency consists of identical coins or bills, all of which share a set amount of value.

The potential of NFTs goes far beyond art; the real advancement lies in the concept of being able to prove, through public code, that something is owned. Everything can be tokenized and become an NFT, and ownership can be proven without relying on intermediaries; manipulation and falsification of data is impossible. To make these concepts clear, consider a few of the following applications of NFTs.

1. NFTs in Art.
2. NFTs in Gaming.
3. NFTs in Sports.
4. NFTs in Music.
5. NFTs in the Metaverse.
6. NFTs in Verification.
7. NFTs in Real Estate.

NFTs in Art

The current primary application of NFTs is in the art space. NFTs are both touted and teased as being monetizable jpegs, and most people know of NFTs only through art projects like CryptoPunks, CryptoKitties, and artists such as Beeple. Really, NFTs represent something unique, verifiable, and stored in a digital format; art merely serves as an easy bridge to attach something very visible to the not-so-visible technology of tokenization.

The first NFT, called Quantum, was an octagon-shaped animation, and NFT art has since grown to $10 billion in volume in Q3 of 2021 (according to DappRadar) and individual pieces have sold for tens of millions.

*This piece of art, representing 13 years of work by the digital artist Beeple, sold for $69.3 million at a Christies auction.

NFTs in Gaming

In the current era of gaming, players work to earn in-game items, such as skins, abilities, levels, and weapons. The items they hold in-game are owned by gaming companies, while monetizing in-game skills is difficult. NFTs and Web3 enable a reimagined digital landscape: platforms in which every in-game item is an NFT. Players can legitimately own what they work for and easily trade and profit off such items.

In NBA 2k, what if in-game players could be collected and traded as NFTs, each reflecting the value of real players and going up or down in price as such? In Fortnite, what if you could actually own your skins and transfer them into your digital gallery in a metaverse? These applications and more have already created massively popular NFT gaming experiences: Axie Infinity, Guild of Guardian, Idle Ciber, the Sandbox, Decentraland. Many more are in the works.

*Age of Rust is one of many new games incorporating NFTs.

NFTs in Sports

NFTs allow athletes to utilize their brand to increase personal earnings and collaborate with fans in new and better ways. Collectibles in sports have historically operated through a few physical verticals; take baseball cards as an example. Baseball cards are largely valued by rarity and player performance. However, the number of a certain baseball card in existence is unknown, and athletes are rarely able to profit off cards and at the very least can't collect royalties.

NFTs offer an alternative option—players minting their own digital cards. Fans who believe in the potential of the athlete can buy the card, and the athlete earns royalties each time the card is traded. The exact number of cards in circulation is known, and value is easier to determine.

NBA Top Shot did exactly this and officially licensed digital NBA collectibles. The project has more than a million users and grossed over $700 million last year. To compare, the combined market value of the entire sports card and memorabilia industry is estimated to be $5—that's valuation, not sales.

Additionally, sports players are increasingly requesting contracts to be paid out in bitcoin—most recently, Odell Beckham Jr. announced that his entire 2021-22 salary would be paid out in crypto. Some stadiums have joined in on the trend and offer crypto as a seat payment option, and the LA Staples Center was recently renamed to the Crypto.com arena in a deal worth $700m over 20 years. Across the board, sports players, teams, and organizations are exploring the ways in which NFTs and cryptocurrency can enhance the player and fan experience.

NFTs in Music

The current model for musicians involves a centralized organization, most traditionally a record label, that pays out recoupable advances and royalties. Musicians must also split money among agents, lawyers, distributors, and other parties. In applying non-fungibility and decentralization to the music industry, individual records and albums, visual materials, tickets, and merchandise can all be tokenized and provide a new model for artist-community interaction.

Fan funding, especially, cuts out layers of intermediaries and enables fans to become minters (co-owners) of music, receive unique access to artists and collections, fund projects, support their favorite creators, and profit off the growth and evolution of such creators. For example, a fan may provide 1% of the funds necessary to launch a track for an up-and-coming artist, and a smart contract can send 1% of all royalties earned off that track directly to the fan.

While these applications certainly have flaws and have yet to develop to a mainstream degree, artists such as Grimes, Shawn Mendes, Kings of Leon, Devon Welsh, Young and Sick, and Steve Aoki have all entered the NFT space, and many more musicians will continue to do so.

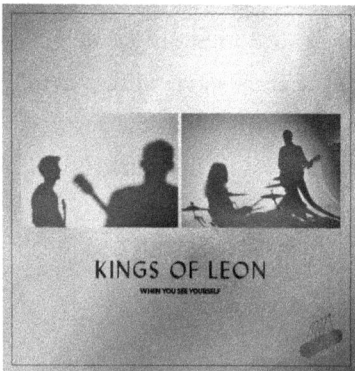

KINGS OF LEON
WHEN YOU SEE YOURSELF

*Kings of Leon was the first band to release an album as an NFT, with each token unlocking special perks including limited-editions vinyls and front row seats. The band has already made millions, and Kings of Leon NFTs are available on opensea.io.

NFTs in the Metaverse

The "metaverse" may be defined in several different ways. Some see it as the moment at which our digital lives become more valuable to us than our real-world lives. Conventional systems think of it as a fully immersive, virtual reality world.

No matter how you define it, non-fungible digitized items play into the concept of a metaverse, since things in digital spaces must be owned just like in real life, and the only feasible route to truly being able to own digital items, as opposed to a central intermediary holding the items, is through NFTs.

As opposed to occupying a plot of land created by a centralized company, NFTs can let you irrevocably own a plot of land in a virtual world created through decentralized means. Just imagine: in the house you build on your digital plot of land in the metaverse, you can frame your NFT art on the wall, and then you can play an immersive game in which you earn an NFT skin, which you can sell to cover the cost of the materials used to build your house. In ways like these and more, virtual worlds are a breeding ground for NFT imagination and implementation.

*A look at The Sandbox (sandbox.game) metaverse. Each plot is owned by a user, not the game, and each piece of land is now worth many tens of thousands of dollars.

NFTs in Verification

NFTs provide an unalterable means of verification; this use case expands across a multitude of industries and applications. Colleges can tokenize degree certificates to prevent fraud; sport organizations can tokenize tickets and collectibles to prevent counterfeits; governments can tokenize personal documents (birth certificate, ID, etc.) to irrefutably prove legitimacy. Any item or organization that benefits from provable history can benefit from non-fungibility.

*A startup called Kred (at nft.kred) offers NFT ticket creation tools to brands.

NFTs in Real Estate

Real Estate is innately non-fungible and transferring Real Estate to the blockchain is not a large jump. The property rights of say, a house, can simply be added to the blockchain, resulting in unalterable and verifiable ownership. In war-torn or unstable countries, applications such as these can prove incredibly useful, and in all real estate markets, NFTs cut out intermediates, save on cost, and increase speed.

*NFTs in real estate goes far beyond personal property. Cryptoisland.com is utilizing DAOs and NFTs to community-own a $35m private island in the Bahamas.

Concept Summary:

NFTs are digital provenance.

Through blockchain, NFTs prove ownership in a digital and transparent manner.

NFTs remove intermediaries, grant immutable digital ownership, and enable decentralized collaboration.

Essential Terms

Account

An account is a pair of public and private keys from which you can control your funds. You typically view your account through an exchange, which provides an ideal trading interface. However, your funds are actually stored on the blockchain, not in your account.

Address

An address, also known as your public key, is a unique collection of numbers and letters that function as an identification code, comparable to a bank account number or an email address. With it, you can carry out transactions on the blockchain. Addresses have round, colorful "logos," that are called address identicons (or, simply, "icons"). These icons allow you to quickly see whether or not you input a correct address.

Airdrop

An airdrop is a marketing tool used by new coins. The team behind a new coin or token will give users the ability to receive the asset for free, typically in exchange for a small task, such as following the company on social media or providing an email address. Airdrops are great for projects, since many new customers get excited about the coin and want to see it rise in value. It's also great for users, since they get the coin for free and can potentially make a lot of money.

However, airdrop scams are common, and many new coins fail, so make sure to do your research to understand what new airdrops good, and what airdrops are aren't. Here are a few sites that provide information about new airdrops:

- aidrops.io
- airdropalert.com
- icomarks.com
- cocoricos.io

Algorithm

An algorithm consists of the mathematical rules that a code or software must follow. Many forms of algorithms are used across the internet, such as those used by social media services to decide which content gets how much exposure. Blockchains and cryptocurrencies use algorithms to perform a variety of tasks.

Bitcoin

Bitcoin was the first cryptocurrency. It was created in 2008 by an individual or group of individuals operating under the name of Satoshi Nakamoto.

Cash

In the world of crypto and investments, cash does not mean keeping literal cash, but rather money that is not invested and rather being held in a digital account.

Confirmed

Transaction confirmation refers to a transaction being confirmed, which means multiple peers in the network have validated the given transaction. Once a transaction has been confirmed, it is permanently stored and viewable in the public ledger.

dApp / DAOs

dApp is short for "decentralized application." Basically, any app that runs on a blockchain (or any other peer-to-peer network) and does not have a centralized owner is considered a dApp. DAO is shorthand for decentralization autonomous organization and refers to any organization that is transparent, owned by a network of distributed participants, and run by programmed rules as opposed to a centralized structure.

Decryption / Encryption

Encryption is the process of converting plain text into coded information through the use of a cipher. The opposite is decryption, which converts coded information into plain text. Decryption in crypto involves turning encrypted data back into plain text.

Digital Commodity

A digital commodity is a digital asset that holds value. Digital commodities do not have to be digital currencies. NFTs, digital art, and anything else that holds value and exists online are digital commodities.

Digital Currency

Digital currencies lie within the realm of digital commodities. Instead of referring to all digital assets, digital currencies refer to all currencies that operate solely online and do not have a physical form.

Digital Signature

Your digital signature is used to confirm that online documents come from you. This is not equivalent to a physical signature. Instead, digital signatures are code generated by an algorithm.

Distributed Ledger

A distributed ledger is a ledger that is stored in many different locations so that transactions can be validated by multiple parties. Blockchain networks use distributed ledgers.

Dolphin / Whales

Crypto holders are classified through a few different animals. Those with extremely large holdings are called whales, while those with moderately sized holdings are called dolphins.

Dump

To dump, or dumping, refers to selling a large amount of cryptocurrency or a large amount of a coin or token being sold. For example, you might say "that coin is dumping," or "I'm dumping this coin."

ERC-20 / ERC-20 Standard

An ERC-20 is one of the many Ethereum token types. Remember, a token is a token because it is built upon another blockchain, while coins are built on their own blockchains. ERC-20 is significant in the world of Ethereum tokens because it is used to define the rules by which all tokens on the Ethereum blockchain function. It can be likened to a security guard; it requires and ensures that all tokens in its vicinity follow that set of rules. The ERC-20 "standard" is the combined list of all the rules. Tokens using the ERC-20 standard can transact between each other and exchange in a more efficient manner.

Ether

Ether is the native cryptocurrency of the Ethereum blockchain. Its ticker symbol is ETH, and to use any currency on the Ethereum blockchain you must pay fees in Ether.

Exchange

A [cryptocurrency] exchange is a marketplace in which cryptocurrencies are traded. Exchanges must be combined with wallets. In wallets, coins can be held through addresses. Exchanges act as an easy intermediary to help users transact.

Fiat

Fiat refers to governmental currencies, such as the US dollar and Euros.

Fintech

Fintech is short for financial technology. Fintech consists of any technology that supports and/or enables financial services. Cryptocurrencies are fintechs, as well as companies such as GoFundMe and PayPal.

Fork / Hard Fork / Soft Fork

A fork is the occurrence of a new blockchain being created from another blockchain. For example, Bitcoin Cash once forked off from Bitcoin. Forks occur when algorithms have a disagreement and split into two different versions. Two kinds of forks exist: A hard fork and a soft fork. A hard fork in a blockchain is a fork that occurs when all the nodes in the network upgrade to a newer version of the blockchain and leave the old version behind. Two paths are then created: the new version and the old version. A soft fork contrasts this by rendering the old network invalid; this results in just one blockchain, not the two that comes as a result of a hard fork.

Fundamental Analysis

Fundamental analysis is analysis of a coin or token through its fundamental metrics. Fundamental metrics looks at economic and financial activity to determine value.

Gas

Gas refers to the fee required to complete transactions on the Ethereum blockchain. Gas fees are given to the miners, who validate blocks and ensure secure networks.

Gwei

Gwei is the denomination (the price-per-unit) used in defining the cost of Etherium gas. You can think of Gwei and Etherium as similar to the penny versus the dollar. 1 ETH equals one billion Gwei. Gwei is used instead of Etherium because seeing that gas fees are 1 Gwei is easier than seeing the fee as 0.0000000001 Ether. That said, gas fees are quite high as of 2022, and for this reason resorting to Ether denominations is currently more applicable, though this will not be the case forever.

Halving

Halving is the process by which the reward for mining bitcoin is cut in half. Bitcoin halving happens every 210,000 blocks, which roughly equates to every 4 years. Halving will happen until the maximum supply of bitcoin has been reached and all 21 million coins are put into circulation.

Hash / Hash Rate

A hash is a function that converts one value into another; a hash in the crypto world converts an input of letters and numbers (a string) into an encrypted output of a fixed size. Basically, hashes help with encryption. "Solving" each hash requires working backwards to solve an extremely complex mathematical problem. The measure by which a computer is judged in terms of its ability to hash is called a hash rate. Put simply, the hash rate is the speed at which a node can perform hashing, and hashing is important in cryptography.

Hot Wallet / Cold Wallet

A hot wallet refers to a cryptocurrency wallet that is connected to the internet. The opposite, cold storage, refers to a wallet that is not connected to the internet. Hot wallets allow for the owner of an account to easily send and receive tokens; however, cold storage is more secure than hot storage.

Initial Coin Offering (ICO)

In order to raise funds and awareness, the creators of a cryptocurrency often put an initial portion of their coin supply up for purchase.

Initial Exchange Offering (IEO)

An IEO is similar to an ICO. Both are initial offerings of coins or tokens used solely within the crypto space. IEOs are coming into fashion as the improved version of ICOs because IEOs allow online crypto trading platforms to directly make the asset tradeable. Basically, IEOs require less effort to invest in and streamline the trading process of an initial offering.

Keys

A key is a random string of characters used by algorithms to encrypt data. Two keys are used for cryptocurrency: a public key and a private key. Both are important to understand and are defined in depth below.

Mining

Mining is the process by which blocks are added to a blockchain through the solving of mathematical problems. Solving these problems takes an extremely large amount of computational power, hence, rewards are provided to those who do the work. People or organizations who use their computational power to mine are known as "miners."

Network

A network, at its core, is an interconnected system. The system within a cryptocurrency network is made up of many nodes (devices) that assist a blockchain in a variety of tasks. So, a crypto network can be thought of as many different computers working together to run a blockchain.

Node

A node is a computer connected to a blockchain's network that assists the blockchain in writing and validating blocks. Some nodes download an entire history of their blockchain; these are called masternodes and perform more tasks than regular nodes. Additionally, nodes are not locked into a specific network. Rather, most nodes can switch to different blockchains practically at will, as is the case with multipool mining.

Peer-to-Peer (P2P) / P2P Networks

A peer-to-peer network involves many computers working with each other to complete tasks. Peer-to-peer networks do not require a central authority and are an integral part of blockchain networks.

Private Key / Public Key

Cryptocurrency users will utilize two keys: a public key and a private key. Both keys are strings of letters and numbers. Once a user initiates their first transaction, a pair of both a public key and a private key is created. The public key is used to receive cryptocurrencies, while the private key allows the user to carry out transactions from their account. Both keys are stored in a crypto wallet.

Protocol

A protocol is a system or procedure that controls how something should be done. Within cryptocurrency, protocols are governing layer of code. For example, a security protocol determines how security should be carried out, a blockchain protocol governs how blockchain acts and operates, and a Bitcoin protocol controls how the Bitcoin network functions.

Pump / Dump

A pump is a rapid upward price movement in a coin or token. A dump is a rapid downward price movement in a coin or token. "To the moon" refers to a massive pump.

Rank / Ranking

Cryptocurrencies are ranked by market cap. Within the ranking system, which may be thought of like a scoreboard, being in the top 10 is equivalent to a badge of honor. You'll often hear people say, "I think this could be a top 10 coin," and similar statements. Bitcoin has held the top spot since inception and is likely to hold that spot for at least another few years. Check out the coin rankings for yourself at any of the following sites:

- coinmarketcap.com
- coingecko.com
- cryptoslate.com

Satoshi Nakamoto

Satoshi Nakamoto is the individual or possibly the group of individuals who created Bitcoin. Not much is known about this mysterious figure, and his anonymity has spawned countless conspiracy theories. While Nakamoto lists himself as a 45-year-old male from Japan on an official peer-to-peer foundations website, he uses British idioms in his emails. Additionally, the timestamps of his works better align with someone based in the US or the UK. Nakamoto currently holds a fortune worth more than $50 billion through holdings of 1.1 million bitcoins.

Seed / Seed Phrase

A seed phrase is interchangeable with a mnemonic phrase. Seed phrases are 12-to-24-word sequences that identify and represent a wallet. With it, you can never lose access to a connected account. If you forget it, there's no way to reset it or get it back. Anyone who has your seed phrase has full access to the connected wallet and cryptocurrency holdings.

Smart Contracts

Smart contracts are an essential part of the cryptocurrency world. A smart contract is a self-executing contract run on code. The terms of the contract, as well as the execution, is directly written into the smart contract and therefore removes the issue of trust for all parties in the transaction. Transactions issued with smart contracts are irreversible and untraceable. These contracts can be used not just for managing cryptocurrency transactions, but also in voting systems, various financial services, information storage, and throughout a multitude of other industries.

Stablecoin

A stablecoin, similar to a pegged currency, is a coin or token that is designed to remain at the same price of a designated asset, typically a government-issued currency. For example, USDT and DAI are two popular stablecoins pegged to the US dollar, meaning that 1 USDT and 1 DAI will each exist in perpetuity as equal to 1 US dollar. Stablecoins experience practically no volatility, typically provide a few percent interest (APY) on holdings per year and are a good place to store money in the crypto ecosystem.

Technical Analysis

Technical analysis is a type of analysis that looks at technical indicators in order to predict price movement. Technical analysts use historical data from charts to make their predictions.

Ticker / Ticker Symbol

A ticker is a sequence of letters that identifies a specific coin or token. All stocks, as well as cryptocurrencies, have ticker symbols. For example, bitcoin is symbolized through BTC and Ethereum through ETH.

Token

While cryptocurrency coins are built upon their own blockchain, cryptocurrency tokens are built upon a non-native blockchain. Many tokens use the Ethereum blockchain, and are thus referred to as tokens, not coins. Token uses are represented under subcategories, the most essential of which are security tokens, platform tokens, utility tokens, and governance tokens. Understanding tokens is an integral part of understanding what exactly you're trading, as well as understanding all uses of digital currencies, and for those reasons we will take a brief look at the token types just mentioned.

- ◆ Security tokens represent legal ownership of an asset, whether digital or physical. The word "security" in security tokens doesn't mean security as in being safe, but rather, "security" refers to any financial instrument that holds value and can be traded. Basically, security tokens represent an investment or asset.

- Utility tokens are built into an existing protocol and can access the services of that protocol. For example, utility tokens are commonly given to investors during an ICO. Then, later on, investors can use their utility tokens as a means of payment on the platform that provided the tokens. The macro definition to keep in mind is that utility tokens can do more than just serve as a means to buy or sell goods and services.

- Governance tokens are used to create and run voting systems for cryptocurrencies that enable functions such as system upgrades.

- Payment (transactional) tokens are used solely to pay for goods and services.

Transaction

A transaction is any exchange between multiple parties. A cryptocurrency transaction involves one party buying a coin or token, and another party selling that coin or token.

Unpermissioned Ledgers

Unpermissioned ledgers are ledgers that have no single owner. The purpose of such a ledger is to allow for all the benefits of decentralization.

Wallet

A wallet is the user interface that you use to manage your account(s). Holdings are not actually stored in wallets, which are accessible through a private and public key, but rather on a blockchain. Coinbase wallet and Exodus are common wallets

General Knowledge Terms

The following terms constitute general knowledge terms that in some way relate to cryptocurrency.

Angel Investors

Angel investors are wealthy individuals looking to invest in and provide funding for start-ups and entrepreneurs.

Black Swan Event

A black swan event is an entirely unexpected occurrence. For example, if you invest in a coin and the next week news comes out that the project received funding from a group of esteemed angel investors, that coin may unexpectedly rise in value.

Brute Force Attack (BFA)

A brute force attack is an attack that simply "brute forces" information, such as a password, by simply trying as many combinations as possible. Advanced brute force attacks can generate millions of combinations per second. BFAs are the reason you can only try to enter a password a few times before you get locked out.

Bug Bounty

A bug bounty is a reward given to a person or people who find errors or vulnerabilities in a computer program or the wider system. Large companies often offer millions of dollars in bug bounty rewards, as identifying vulnerabilities could prevent hacks.

Cipher

A cipher is the name given to any algorithm, offline or online, that encrypts and decrypts information.

End Users

End users are the people who use the final version of a product (e.g., at the end of the creation process). So, developers and beta testers are not end users, while a consumer who buys a product from a major retailer is an end user.

Escrow

Escrow refers to a third party that holds funds during a transaction. This third party should be unbiased and makes sure that both parties stick to an agreed-upon deal.

Initial Public Offering (IPO) / Direct Listing

An IPO is the process by which a company can become publicly traded on a stock exchange. Crypto coins or tokens can have an IEO or an ICO, but not an IPO. However, some centralized crypto companies grow large enough that they have an IPO. A company can also become listed on a stock exchange through a direct listing, and both methods have a common end result. Coinbase was the largest IPO of a crypto exchange.

Know Your Customer (KYC)

Know your customer guidelines require that finance professionals get about the identity of their customers. KYC procedures fit within the wider AML (Anti-Money Laundering) policies.

Latency

Latency is the delay between when a transaction is submitted and when the network recognizes the transaction. Latency may be thought of as lag.

Liquidity

Liquidity is how easily an asset can be bought or sold. For example, stocks and cryptocurrency are extremely liquid, since they can be bought or sold at a moment's notice. However, assets such as real estate and priceless art may be less liquid, since to sell them requires significant time, effort, and money.

Roadmap

A roadmap is a plan that an organization publishes regarding its long-term goals and important benchmarks.

SEC

The Securities and Exchange Commission, (SEC) is a governmental agency responsible for regulating security markets, such as the stock market.

Security Audit

A security audit is an analysis of how safe a system is against attacks or technical failures. Companies often perform security audits in order to improve their security measures.

Sentiment

Sentiment describes the attitude of a person or group of people towards something, perhaps a company, person, market, or asset. In a nutshell, sentiment is emotion.

Unit of Account

Units of accounts measure asset value. US dollars, as well as other governmental currencies, are units of account because they evaluate things that hold value. For example, if you buy a loaf of bread for $5, the unit of account (the money you used to buy the bread) is being measured against the value of the bread.

User Interface (UI)

The UI is the interface through which users interact with software. Each website you visit is displaying its user interface, which lets you interact with the website code.

Whitelist

A whitelist is a list of approved items or participants, such as at an event. The opposite is blacklist, which is a list of banned items or participants.

White Paper

A whitepaper is a document that startup companies give to potential investors that reveals information about the company plans.

Blockchain Terms

Blockchain is the technology from which cryptocurrency arose, playing the same part as the semiconductor did to Silicon Valley. An overview of blockchain is on page 14, while this section consists of several dozen essential blockchain terms.

Atomic Swap
An atomic swap is a smart contract technology that allows users to exchange two different coins for each other without a third-party intermediary, usually an exchange, and without needing to buy or sell. Centralized exchanges such as Coinbase cannot perform atomic swaps. Instead, decentralized exchanges allow for atomic swaps and give full control to users.

Block
A block, contextually used as part of a blockchain, is a data structure that contains information about transactions. Such data usually includes the amount and time of the transaction, as well as the addresses involved.

Block Height
The block height is the number of blocks in a blockchain. Height 0 is the first block, (referred to as the genesis block) height 1 is the second block, and so on. Currently, the block height of Bitcoin is past half a million. The "block generation time" of Bitcoin is currently around 10 minutes, meaning that one new block is added to the Bitcoin blockchain approximately every 10 minutes.

Block Reward
A block reward refers to the number of coins a miner can earn per each successfully mined and validated block.

Blockchain
A blockchain is a type of database that organizes each list of transactions (referred to as blocks) into chains, hence the name. Blockchain networks use DLT (decentralized ledger technology) and peer-to-peer networks to create decentralized, anonymous, and secure networks.

Chain Linking

Chain linking refers to the process of one cryptocurrency being transferred to another. Since each currency has its own blockchain and the transaction must be recorded on both blockchains, the chains of data on each blockchain is linked to complete the transaction.

Chain Split

A chain split refers to the same event as fork.

Consensus

When a transaction is made on a blockchain network, many different nodes in the blockchain must validate it and reach a consensus on whether or not the transaction is valid. Consensus simply means majority opinion.

Consortium Blockchain

A consortium blockchain is a blockchain network that is privately owned by multiple parties.

Cryptographic Hash Function

A cryptographic hash function is a certain process that happens within nodes. Each node will convert a transaction (or any other input) into an encrypted string made up of letters and numbers that registers the place of the transaction in the blockchain.

Difficulty

Difficulty, in the crypto space, refers to the cost of mining. The difficulty can change from moment-to-moment based on demand and supply.

Full Node

A full node is a node that downloads and contains the entire history of a blockchain in order to completely enforce its rules without requiring the assistance of other nodes.

Gas Limit

When a transaction is made on a network, a user can set the limit they wish to pay in gas fees. Manually setting higher gas limits will result in transactions being fulfilled quicker, since the reward is higher. Gas price is typically set automatically to the going market rate.

Genesis Block

The genesis block is the first block in a blockchain.

Group Mining / Mining Pools

Group mining refers to groups of people or entities who combine their computational power in order to mine together and split the rewards. Group mining is synonymous with mining pools.

Hash Rate

The hash rate is the speed at which a node can perform hashing. Hashing essentially encrypts or decrypts data.

Keys

A key is a random string of characters that are used by algorithms to encrypt data. Both public keys and private keys are used in cryptocurrency.

Key Pairs

A key pair is a combination of a public and private key. All wallets have a unique key pair attached to them.

Layer 2

Layer 2 is a secondary framework or protocol built on top of a blockchain system. Most layer 2 protocols are designed to boost blockchain scalability.

Ledger

A blockchain ledger stores data about all financial transactions made on a given blockchain. Cryptocurrencies use public ledgers, meaning all transactions performed with that cryptocurrency are publicly viewable through a ledger. Refer to the Blockchain section for more information on public ledgers.

Lightning Network
A lightning network is a type of secondary layer on top of a blockchain system. Lightening networks enable faster transactions.

Mainnet
The mainnet constitutes the main network of a blockchain. Transactions take place on the distributed ledger, not the mainnet.

Mainnet Swap
A mainnet swap occurs when a coin moves from one mainnet to another.

Masternode
Masternodes have more roles than a regular node in a blockchain network, such as enabling specific services.

Mempool
A mempool is a node's holding area for pending (unvalidated) transactions.

Merkle Tree
A merkle tree is a way in which data is structured. The Merkle structure gives off the visual appearance of being a tree, and Merkle Trees are synonymous with Hash Trees.

Merged Mining
Merged mining is the act of mining multiple cryptocurrencies at the same time.

Mining
Mining is the process by which blocks are added to a blockchain through working backward to solve mathematical problems. Solving these problems takes an extremely large amount of computational power. To make up for this cost and

incentivize activity, rewards are provided to those who do the work. The people or organizations who use their computational power to mine are the miners.

Mining Contract

A mining contract is a contract that involves one party loaning (basically, renting) their computational power (hashing power) to another party. The purchaser pays an upfront fee in return for the rewards generated by the rented hashing power.

Mining Farm

A mining farm is a collection of multiple miners, typically a large group, who manage a large data center or warehouse devoted specifically to mining cryptocurrencies. Dave Carlson manages one of the largest mining farms in the world. His monthly expenses are more than $1 million worth of electricity, and although specific numbers are not public, it's estimated that Carlson at some point mined up to 200 bitcoins per day. Please keep in mind that mining farms are illegal in many places around the world.

Mining Pool

A mining pool is a group of miners who combine their computational power to earn rewards faster. The rewards are then split across the group relative to the amount of power contributed. A mining pool is better than individual mining for those with relatively little computing power because rewards are distributed more often. So, as opposed to a small miner earning one big dump of $1000 worth of cryptocurrency every two years, that miner may join a pool and earn $1.37 per day.

Multipool Mining

Multipool mining is the event of miners moving from one cryptocurrency to another based upon the rewards offered. Pools offering multipool mining automatically switch between coins. Rewards are then distributed to the miners who contributed computational power.

Multisignature

Some wallets require multiple parties to authorize and validate transactions before such transactions are added to the public ledger, hence requiring a multisignature. Multisignature addresses and multisignature wallets require multisignatures.

Network

A network, at its core, is an interconnected system. The system within a cryptocurrency network is made up of many nodes that assist the blockchain in a variety of tasks.

Node

A node is a computer connected to a blockchain's network that assists the blockchain in writing and validating blocks. Some nodes download an entire history of their blockchain; these are called masternodes and perform more tasks than regular nodes. Additionally, nodes are not locked into a specific network. Rather, most nodes can switch to different blockchains practically at will, as is the case with multipool mining.

Nonce

A nonce is an arbitrary number used just once to verify a cryptographic transaction. The nonce is the number crypto miners are looking to find by solving mathematical equations.

Oracles

Smart contracts within a blockchain can only be externally reached through oracle programs. Oracles send data to and from smart contracts and external sources as required; you may think of them as performing the same tasks as messenger RNAs in the human body.

Order Book

An order book is a list of open buys and sell orders for an asset on an exchange. Any order that has not been filled is an open order in an order book.

Peer-to-Peer (P2P) / P2P Networks

A peer-to-peer network involves many computers working with each other to complete tasks. Peer-to-peer networks do not require a central authority and are an integral part of blockchain networks.

Permissioned Blockchain / Ledgers

Most crypto blockchain networks are public; this means that anyone can mine the blockchain and assist in adding blocks. An alternative to this system is a permissioned blockchain, which includes a corresponding permissioned ledger. Within permissioned networks, an access control layer is built on top of the normal blockchain that controls, in a nutshell, who can do what.

Private Key / Public Key

Cryptocurrency users will utilize two keys: a public key and a private key. Both keys are strings of letters and numbers. Once a user initiates their first transaction, a pair of both a public key and a private key is created. The public key is used to receive cryptocurrencies, while the private key allows the user to carry out transactions from their account. Both keys are stored in a crypto wallet.

Public Blockchain / Private Blockchain

A public blockchain is an open network that allows any computer to participate in verifying transactions. On the other hand, private blockchains regulate who has access to the network and who may participate in the network.

Ring Signature

A ring signature is a digital signature (it can also be thought of as an encryption process) that allows both the giver and the receiver to remain anonymous by giving nodes within the network the power to approve transactions without identifying which node requested the transaction, hence removing any digital trail between the two nodes and keeping the keys and the identity of the sender and receiver private.

Seed / Seed Phrase

A seed phrase is interchangeable with a mnemonic phrase. Seed phrases are 12-to-24-word sequences that identify and represent a wallet. With it, you can never lose access to a connected account. If you forget it, there's no way to reset it or get it back. Anyone who has your seed phrase has full access to the connected wallet and cryptocurrency holdings.

Smart Contracts

Smart contracts are an essential part of the cryptocurrency world. A smart contract is a self-executing contract run on code. The terms of the contract, as well as the execution, is directly written into the smart contract and therefore removes the issue of trust for all parties in the transaction. Transactions issued with smart contracts are irreversible and untraceable. These contracts can be used not just for managing cryptocurrency transactions, but also in voting systems, various financial services, information storage, and throughout a multitude of other industries.

Staking Pool

A staking pool is a group in which stakeholders combine their computational staking power in order to increase their ability to successfully validate a new block. A block reward is earned each time a block is validated, and rewards are thereafter distributed in accordance to contribution.

Timestamp

A timestamp is part of every block within a blockchain. Each timestamp contains the exact moment in which its block was mined and validated. This assists in confirming that blocks do not get tampered with.

Unconfirmed

Unconfirmed transactions are transactions that have not yet been verified and put onto a blockchain.

Unpermissioned Ledgers

Unpermissioned ledgers are ledgers that have no single owner. The purpose of such a ledger is to allow for all the benefits of decentralization, most notably transparency, efficiency, and security.

Yield Farming

Yield farming puts crypto assets to work in order to generate returns. Yield farming involves users lending funds to others through smart contracts. In return for lended funds, the lender earnings rewards in crypto. Yield farming has recently become quite popular.

Trading

Trading cryptocurrency is a massive topic, and this section is my best attempt to ensure your success as an investor in cryptocurrencies. Here is a look at everything coming up:

- How to Pick Crypto Investments

- Technical Analysis

- Fundamental Analysis

- Hype Analysis

- Investing Rules

- Charts

- Patterns

- Indicators & Oscillators

- Investing Strategy

- Trading Terms

How to Pick Crypto Investments

Picking good investments is always a guess. Each time you purchase an investment, someone else is selling it, and therein lies the basic cognitive clash of trading. No matter how easy it may seem, investing in any asset class is never a sure thing, and this has been proven time and time again across the stock market, real estate, bonds, private equity, and so on.

Many traders have discovered this fact in crypto throughout the past few years despite the magical returns of crypto markets over the past decade and entering cryptocurrency investing with the preconceived notion that it is easy has proven to be dangerous. On average, investors in most major asset classes tend to underperform the market, and eToro's study of day traders found that 80% lost money over a 12-month period. Most traders become part of this statistic through over-trading and a lack of proper investing knowledge, and all crypto investors, not just day traders, face these same pitfalls.

These next few pages break down the three means of analysis, as well as trading rules, all to firmly implant a groundwork that enables profitability. The importance of such information cannot be stressed enough–investing right is just as glorious as investing wrong is painful. Take your time, learn the process, and set yourself up for success.

Technical Analysis

Technical analysis is the discipline by which future movements of securities, currency pairs, and cryptocurrencies are discerned from historical patterns. In a nutshell, technical analysis is the investment process of using history to predict the future. History, in turn, is analyzed through the spectrum of patterns and indicators within charts. Technical analysis is backed by a few select premises, called the key three, which collectively dictate the assumptions behind technical analysis. Everything related to technical analysis, such as indicators, analyzing charts, and even the entire basis of buying and selling assets based on historical events, is based upon these statements.

History Tends to Repeat Itself

While the idea that history tends to repeat itself may sound self-explanatory, it is actually quite a novel concept. No rules require the prices of investments must act a certain way, and no inherent intelligence correlates historical movement to current and future movement. However, the entire basis of technical analysis requires that history does repeat itself, because if history repeats itself, history can be predicted, and if history can be predicted, money can be made. Assuming the above, tendencies to repeat historical price action must be due to outside influences, namely, the investors themselves. Much of this can be traced back to investor psychology and self-predicting patterns, while much of the rest is due to institutional investment patterns.

The Market Discounts Everything

The idea that the market discounts everything, alternatively phrased as "market action discounts everything," is a part of the Efficient Market Hypothesis (EMH). The EMH states that prices (within our context, the prices of cryptocurrencies) reflect all available information. Various versions of this theory exist, which are thought of as weak, strong, and everything in-between. Since the crypto market is highly volatile and somewhat more trend-based than other security markets, it is less of an efficient market (relative to most) because prices and price variability may not accurately reflect true value. For example, when Elon Musk tweets about a small-cap crypto and the price increases fivefold, the price increase did not represent an efficient market because the true value of the crypto didn't change, but the price did.

The idea that the market is not completely efficient opens up the possibility of undervalued prices. Technical analysis, in part, aims to identify discounted price (relative to true value) through technical means. Technical analysts care more about what is most likely to happen given historical movements than what may happen based on information already priced into the security, including trading based on earnings, trends, and hype.

Prices Move in Trends

Trends are an exceedingly important concept to technical analysts. The entire point of charting price movements is rendered illegitimate unless one assumes that prices move in trends and that trends are more likely to continue than reverse.

Fundamental Analysis

Fundamental analysis, another means of choosing investments, revolves around analyzing the true value of an asset through valuation techniques that include overall economic analysis, industry and sector analysis, and analysis of financial data.

Since fundamental analysis relies purely on publicly available data, investors can find investments through either a top-down or bottom-up approach. In a top-down approach, the health and direction of the economy are first considered, followed by each sector, and finally each asset. Investors select the best of each stage and funnel down to find undervalued opportunities.

This concept can be taken from the stock market and applied to cryptocurrency fundamental analysis by first researching the overall health of the cryptocurrency market and then identifying undervalued segments. From there, the most undervalued companies and projects may be sought out. An example goes as follows:

I think the cryptocurrency market will go up in value based on current adoption trends.

$\vee$

Within the cryptocurrency market, I have data showing that metaverse related projects are most likely to go up in value since more and more companies and people I known are getting involved in AR and VR.

$\vee$

I've looked at all the crypto projects within the metaverse space, and upon further research, I have found the most undervalued company to be The Sandbox (SAND).

In this way, one may use a top-down analysis to identify the best investments. However, the nature of the process requires a significant amount of time spent sorting through all sorts of data, first for the overall market, then through various sectors, and finally through all the currencies in that sector.

The opposite approach, called a bottom-up analysis, first analyzes individual assets. This can work because an individual investment will often outperform its overall industry or sector, so relying on sector data ignores the value of overperforming projects within underperforming sectors.

The core idea of cryptocurrency fundamental analysis asserts that coins and tokens have an intrinsic value that should reflect upon price. Therefore, one can conclude that any price under this true value (true value not being stagnant and changing as new information becomes public) renders it undervalued, and any price over the intrinsic value renders it overbought and a sell. Although the concept of value and identifying value may sound like an exact science (e.g., the true value of this crypto is $20 and it is trading at $15, easy $5 potential gain), fundamental analysis is quite a speculative matter since true value itself is a matter of speculation. The central idea is that all crypto investments are either under valued or overvalued, and your job as a fundamental analyst is to identify the most undervalued, buy, and sell once properly or overly valued.

Fundamental analysts conduct research through a wide range of sources, but most information can be assembled through the following channels:

Teams
Whitepapers
Events
Competition
Utility
Sentiment Analysis
Market Cap
Activity
Volatility
Supply Mechanisms

These ten elements of fundamental analysis will conclude the fundamental analysis section and preclude the final means of identifying good crypto investments, which is hype analysis.

Teams

Each coin or token has a person or a team behind it that aims to provide a service, solve a problem, or in some other way provide utility and value. Information on the team behind potential investments is often a great indicator of long-term success. This applies to a lesser degree if a trade is meant to be executed in a short period; still, even then, performing due diligence on a team allows for a greater understanding of an overall situation and provides important context to all investment decisions.

Take Storj, a project that aims to create decentralized cloud storage. The team consists of eighty reputable experts from varying backgrounds, and the CEO, Ben Golub, is a professor at Northwestern University and previously taught at Harvard. Additionally, major companies such as Google have put money into the team and project. While nothing here guarantees success, a solid foundation does massively increase the chances of long-term innovation and success.

Some cryptocurrencies aren't run by a stand-alone group of developers, but rather by organizations. Cardano, an example of such, is run by three companies: IOHK, Emergo, and the Cardano Foundation, which, in turn, are all managed by reputable industry leaders. ADA, in fact, is up 4,000% over the past few years.

Projects with people and organizations like ADA and STORJ, as well as many others, are the gold mines of the crypto space. The projects that survive in the long run and continue to build utility and create value are the projects that are likely to deliver massive and sustained growth, and such projects are founded upon competent and dedicated teams. To research the team behind a project you're interested in, just search around, check out the project's website, and look into the background of executives.

White papers

A white paper is an informational report issued by an organization about a given product, service, or general idea. White papers present information about the idea of the project and provide a timetable for future events. Generally, this helps readers understand a problem, figure out how the creators of the paper aim to solve that problem, and form a conclusive opinion about the project viability.

Three types of white papers frequent the business space: the "backgrounder," which explains the background of a product, service, or idea, and provides technical, education-focused information that sells the reader on the concept. A second type of white paper is a "numbered list" that displays content in digestible, number-oriented format. For example, "10 use cases for coin HL" or "10 reasons token CM will dominate the market." A final type is a "problem/solution" white paper, which defines the problem that the product, service, or idea aims to solve and provides the implemented solution.

White papers are used within the crypto space to explain novel concepts and the technicalities, vision, and plans surrounding a given project. All professional crypto projects will have a white paper, typically found on their website, and these reports will give you a better understanding of a given project than practically any other single source of accessible information.

Here are a few websites that store crypto white papers.

- ◆ allcryptowhitepapers.com
- ◆ cryptorating.eu/whitepapers
- ◆ coindesk.com/tag/white-papers

Events

A great way to analyze the potential of a coin or token, whether in the short term, midterm, or long term, is through an understanding of upcoming events. Popular crypto event calendars are below.

- ◆ CoinMarketCal
  coinmarketcal.com/en

- ◆ CoinEvents
  coinevents.co

- ◆ CoinsCalendar
  coinscalendar.com

These websites list all upcoming launches, partnerships, airdrops, forks, swaps, and other notable events for most cryptocurrencies. Simply, the amount of upcoming coin events, not to mention the quality, can tell you a lot about a project. In turn, reading into each event provides a step forward in terms of understanding how a project plans to evolve over time.

Since cryptocurrencies do not form a fully efficient market, events occurring a few weeks or months out are often not fully priced into the asset and therefore represent a separation from true value. Trading solely based on this information is quite risky (and not recommended), and if done at all should focus on events occurring at least a month or more out from the current date. Even then, risk is involved, because if enough traders buy early enough, with the intent of dumping the day of an event after an assumed pump, the price can instead crash, regardless of the event's outcome.

So, weigh all of these event-related factors while considering how or whether to invest in a coin or token and, regardless, make sure to stay in the loop on events happening across the market and within the realm of the assets you hold.

Competition

As within all aspects of business, competition is a must for understanding the relative situation of a company in a given market. Within the highly volatile crypto market, this applies to an even greater degree. There is plenty of space within niches of the crypto market for multiple companies, and this falls back to the Big-Brother concept, which dictates that projects offering a small twist off of another larger project often perform exceedingly well despite established competition.

Examples:

♦ USD Coin positioned itself as an improved version of Tether.

♦ PancakeSwap offers lower fees than Uniswap.

♦ Shiba Inu benefited from the Dogecoin escapade.

♦ Cardano and Solana grew by being viewed as improved versions of Ethereum.

♦ The Sandbox leveraged the hype created by Decentraland.

Utility

Utility within a coin or token is one of the most important aspects of due diligence, since understanding the current and long-term purpose driving a coin or token allows for a much clearer analysis of potential. Coins and tokens with utility have real, practical uses—they don't just exist, but rather solve a problem or offer a service. Coins with the most functional use cases are likely to succeed as opposed to those without continued purpose, use cases, and innovation. Consider the following case studies:

- Bitcoin (BTC) serves as a reliable and long-term store of value, akin to "digital gold."

- Ethereum (ETH) allows dApp and Smart Contracts to be created on top of the Ethereum blockchain.

- Storj (STORJ) can be used to store data in the cloud in a decentralized manner, similar to Google Drive and Dropbox.

- IOTA (IOTA) offers completely free transactions to be used for small, daily payments.

- Basic Attention Token (BAT) is used within the Brave browser to earn rewards and send tips to creators.

- Golem (GNT) is a global supercomputer that offers rentable computing resources in exchange for GNT tokens.

All of these coins have real, practical utility, and projects that have utility and work on constantly improving utility are much more likely to be successful over the long run. Make sure to consider how teams, as mentioned hitherto, play into utility.

Sentiment Analysis

Sentiment analysis is the art of figuring out what other people think. Understanding sentiment towards a person, brand, coin, token, trend, etc., is useful information because social momentum often predicts trends.

Today, software can analyze social media websites and the wider internet for sentiment (for example, identifying the amount of positive versus negative words in tweets mentioning "Bitcoin" in the past 24 hours) and bundle this information into streamlined workflows. You can also do your own research by combing through websites or simply reading article titles.

Below are several tools (all of which are free) that can be used to understand sentiment in the entire market, the entire investment community, or individual assets.

- Crypto Fear & Greed Index
  alternative.me/crypto/fear-and-greed-index/

- Bulls & Bears Index
  augmento.ai/Bitcoin-sentiment

- Santiment
  app.santiment.net

Market Cap

Market cap, shorthand for market capitalization, represents the total value of a cryptocurrency. Market cap provides information on volatility, potential upside, and a host of other factors that influence investment decisions.

To find the market cap of a coin or token, multiply the price by the total number of units. For example, a cryptocurrency with a circulating supply of 1,000,000 coins and a price of $10 per coin has a market cap of $10 million. Here are some other market cap equations:

Helium (HNT)
$18 (price) x 77,995,503 (supply) = $1,403,919,054 (market cap)

Binance Coin (BNB)
$475 x 154,532,785 = $73,403,072,875

Keep in mind that both sides of the equation (the number of units and price) are arbitrary without the other.

Activity Metrics

Activity metrics assist in determining usage, as well as how that usage manifests itself. Three main metrics embody activity: volume, active addresses, and volatility.

The first is trading volume, typically just called "volume." Volume is the number of coins or tokens traded within a specified time frame. Through an understanding of volume, other information about a coin, such as popularity, volatility, utility, and so on, can be better understood. Below are a few sites that provide easy and free information about volume:

- ◆ CoinMarketCap
  coinmarketcap.com

- ◆ CoinGecko
  coingecko.com

- ◆ Yahoo Finance Crypto
  finance.yahoo.com

The second metric is active addresses, which is the number of unique addresses that participate in one or more successful transactions within a given timeframe and given parameters defining "active." It can be thought of as the number of people actively trading throughout a crypto ecosphere, regardless of how much they trade.

Understanding active addresses relative to historical data plays a useful part of understanding the overall adoption trends of a given asset. When researching active addresses, choose relevant parameters of activity, such as activity in addresses with a balance of over $1 million (to see whether large accounts are buying or selling). Check out the number of active addresses on the Bitcoin network here:

- ◆ Glassnode Studio
  studio.glassnode.com/metrics?a=BTC&m=addresses.ActiveCount

The final activity metric is volatility. Volatility is why many people don't get into the crypto market and also why many do; it both creates riches and bankrupts, and in large part has furthered the stigma surrounding cryptocurrencies. Volatility is a measure of deviation: how fast, how often, and how much prices vary. Put simply, it is the size of change. Volatility is typically calculated through standard deviation, and unusual volatility activity usually predates breakouts, whether to the upside or downside.

Different asset classes are known for certain levels of volatility, and this, more often than not, is why or why not an investor gets into a certain investment. Here is a look at several different asset classes:

> Bonds
> Cash
> Cryptocurrencies
> Real Estate
> Stocks

Given the above list, I'll rearrange these asset classes according to volatility (least to most).[1]

> Cash
> Bonds
> Real Estate
> Stocks
> Cryptocurrencies

So, cash has a very low measure of volatility and people who hold a large percentage of their capital in cash are generally risk averse. Real estate is more volatile than cash, so people who get into real estate must be a little more comfortable with risk. The stock market (especially certain areas of the market, such as penny stocks and options) carries higher levels of risk compared to real estate, while cryptocurrency holds the top spot.

---

[1] Consider the following a generalized list of macro market volatility which excludes asset-specific volatility.

You may notice that if the list was rearranged to measure average return from least to greatest, it would stay the same. This is because returns generally correlate with risk, and a principal measure of perceived risk is volatility. Risk-averse investors generally perform much worse than risk-pro investors throughout industry and economic booms, but often fare better throughout recessions and market reversals.

Overall, understanding the volatility and risk of certain investments aids in developing a long-term strategy that suits your investing style and goals. Some people are comfortable with more volatility, others with less, and either way, that's fine; just do you and do your research and hodl.[2]

---

[2] HODL is a common term used in crypto that's a variation of the word "hold" and means the same. I include this to point out that in volatile markets, it's often better to hold for the long term than to make short-term sell decisions.

Supply Mechanisms

Supply mechanisms are the processes by which supply is defined, added, and removed from circulation. In terms of crypto, supply relates to the number of coins or tokens that exist and will exist, as well as how those coins or tokens can be added or removed from circulation. The following pages analyze the concepts of maximum supply and circulating supply and thereafter examine several different types of supply mechanisms, all in order to arrive at a full understanding of how supply mechanics affect investment decisions.

Maximum Supply

The maximum supply is the max number of coins that can ever exist for a cryptocurrency. The maximum supply or the lack of such is pre-set, the most notable example being Bitcoin's 21-million-coin limit. Some coins, like bitcoin, add more coins into the network over time until the maximum supply is reached, while others begin at their maximum supply and others still have no maximum supply. Once a maximum supply is hit, no more coins will ever be procured. Fixed supply coins reach that limit through an "issuance rate" which defines the influx of new coins and typically decreases over time. Contrary to this process, some coins, including Ethereum (ETH), have a set issuance rate and no maximum supply.

To fully understand a cryptocurrency, you may want to check out its maximum supply, as well as its circulating supply. This can be done through popular crypto websites such as coinmarketcap.com and coingecko.com. More information about the tokenomics of a cryptocurrency can usually be found on the project website.

Circulating Supply

The circulating supply is the total number of publicly available coins or tokens. In some cases, such as that of bitcoin, the circulating supply will increase until the maximum supply of 21 million coins is reached. In other cases, the number of circulating coins goes down, often through the process of burning, and thus the intrinsic value of the asset must increase (assuming all other variables are constant) since fewer and fewer will be available. So, the circulating supply is the current number of tradable coins, and the number of tradable coins can either increase or decrease over time.

Fixed Supply - Deflationary Assets

Fixed supply cryptocurrencies algorithmically limit coin supply. Bitcoin is a fixed supply asset since no additional coins can possibly be created once 21 million have been put into circulation. Currently, nearly 90% of bitcoins have been mined, and around 0.5% of the total supply is being lost per year. As a result of halving (covered shortly), Bitcoin will hit its maximum supply around 2140. Many other cryptocurrencies (sourced from cryptoli.st) such as Binance Coin (BNB), Cardano (ADA), Litecoin (LTC), and ChainLink (LINK) operate with similar fixed supplies.

The most apparent benefit of the fixed supply model is that these systems are deflationary. Deflationary assets are assets in which the total supply decreases over time, and therefore each unit increases in value. To illustrate this, say you're stranded on a desert island with ten other people, and each person has one bottle of water. As people drink their water, the total supply of one hundred bottles can only decrease. This makes the water a deflationary asset. As the total supply decreases, each bottle becomes worth increasingly more. Say, now, there are only twenty water bottles left. Each of the twenty water bottles is worth as much as five water bottles were once worth since the total supply has decreased by a factor of five.

In this way, long-term holders of deflationary assets experience increases in the value of their holdings because the fundamental value relative to the whole has increased. For example, at the start of the water-bottle simulation, 1 bottle out of 100 was 1% of the total supply, while by the end 1 out of 20 was 5% of the total supply, making each bottle worth 500% more. In this way, a fixed supply and deflationary model, much like digital gold, will increase the fundamental value of each coin or token over time and create value through scarcity.

Unlimited Supply – Inflationary Assets

Each unit of a deflationary asset increases in value relative to the total supply over time as the total supply goes down. With inflationary assets, new money enters the total circulating supply and causes all the other money to lose value. Return to the island scenario in which ten people have ten water bottles each. Assume that our stranded island-goers are discovered, and a plane will fly over the island and deliver twenty bottles of water per day until the group can be rescued. Each person will then receive two water bottles per day, equivalent to 20% of their total supply. In 30 days, the total supply (ignoring drunk water) will be at seven hundred water bottles, meaning each bottle accounts for 0.14% of the total supply as opposed to the original 1%. This is a 7x decrease in value and reflects the effect of sustained inflation.

The same principal transfers over into securities and cryptocurrencies; many coins have an unlimited supply and experience inflation as a result. Popular coins using an unlimited supply model are Ethereum (ETH), Dogecoin (DOGE), and numerous others.

All that said, an unlimited supply model is not an innately destructive force because of inflation–at least in terms of value, though surely not for our island-goers. Consider Ethereum (ETH), which has an unlimited supply. Eighteen million Ether are mined per year, which is a set constant. Given an infinite amount of time, an infinite number of coins could be produced. However, since the eighteen million additions remain the same while the total supply increases, inflation over time must decrease. To visualize this, if 250 million Ethereum coins exist, inflation is at 7.2%, since 18/250 is 0.072. Yet, in ten years, when 180 million new coins have been minted and the total supply is at 430 million, the same number of coins (eighteen million) are produced, bringing inflation down to 4%. Another ten years down the road and inflation is at 2.9%, and twenty years after that, it is down to 1.8%. In this way, inflation decreases over time. So, while inflation still certainly exists for coins with unlimited supplies, the rate of inflation decreases over time. Additionally, a small amount of inflation is good for the economy, as it forces people to spend or otherwise use money for that money not to lose purchasing power.

So, while most cryptocurrencies have a limited supply and most investors like the idea of deflationary assets, neither the limited nor unlimited models are superior to the alternative. As always, you should do your research and understand what you're getting into, and while supply mechanisms should certainly factor into your decision, they should not be the deciding factor.

Burning

The term "burned" refers to coins being permanently removed from circulation. Burning is a supply mechanism that enables coins to be taken out of circulation, hence acting as a deflationary tool, and thus increasing the value of each other coin in the network, much like company buy backs in the stock market (in which shares are taken off-market).

Burning can be done in several different ways: the most popular simply involves sending a certain amount of the supply to an inaccessible wallet, which is called an "eater address." In this case, while the tokens haven't technically been removed from the total supply, the available supply has effectively gone down. Currently, around 3 million bitcoins (200+ billion of value) have been lost through this process.

Tokens can also be burned through burn functions in the governance protocols, but the far more popular option is through the mentioned eater addresses. As with halving, (immediately below) scarcity creates value, and burning, in theory, increases scarcity and therefore value.

Halving

Halving is a supply mechanism that governs the rate at which coins are added to a fixed supply cryptocurrency. The idea and process were popularized by Bitcoin, which halves every 4 years. Halving is set in motion by a programmed reduction of block rewards, which are the rewards given to the miners that process and validate transactions in a given blockchain network.

From 2016 to 2020, the computers (called the nodes) in the Bitcoin network collectively earned 12.5 Bitcoin every 10 minutes, and that was the number of bitcoins entering circulation. However, following May 11th, 2020, the rewards dropped to 6.25 bitcoin per the same timeframe. In this way, for every 210,000 blocks mined, which equates to roughly every four years, the block rewards will continue to halve until the max limit of 21 million coins is reached around the year 2140.

Thus, halving increases the value of bitcoin by decreasing supply while not altering demand. Scarcity, as mentioned, drives value, and limited supply combined with growing demand creates greater and greater scarcity. For this reason, halving has historically driven the price of bitcoin up and will likely be a long-term growth catalyst.

Hype Analysis

Hype analysis is not a commonly used term in the wider crypto world, but it is a term that adequately describes the phenomenon that is analyzing real-world "hype" trends. Perhaps to a greater extent than any other sizable investment vehicles, the crypto market is driven by hype and trends.

Elon Musk may be the prime example of this, as his tweets about cryptocurrencies are notorious for influencing the price of the subject, whether in a positive or negative manner. Musk once tweeted just the word "Doge," and the price of Dogecoin (DOGE) proceeded to move from $0.036 to $0.082 over the following 5 days, a 220% gain.

While this is mostly unfounded, subcategories within the crypto market, such as DeFi, FinTech, Gaming Coins, Web 3.0, and numerous others, often blow up all at once and cause most of the coins within such areas to experience massive, positive surges. In this way, and others, trading on trends and hype is a strategy that, historically speaking and only if done right, is sound. While I don't necessarily advise this, if it is done right, the sky is the limit.

Trends usually start in social and sentiment-based settings, and this should be the emphasized space in relation to sourcing trends. Such an activity is really just predicting social momentum, and social momentum largely begins on social platforms. Social platforms, in turn, spread information in large part through sentiment (emotion) as opposed to logic-based analysis. The fundamental value behind trends does usually exist, but it often gets blown out of proportion by sentiment.

In recent years, the zenith of this has been the subreddit WallStreetBets and various WSB-induced short squeezes, as well as Elon Musk's tweets in the crypto space. In both cases, trends originated online, were based upon some degree of fundamental value, and were then blown out of proportion by social momentum.

So, the question of sourcing trends is the question of identifying social momentum before it happens. Insight of this kind most often comes by keeping your ear to the ground across crypto communities and influencers on multiple platforms (mainly YouTube, Twitter, Reddit, Discord, Instagram, and TikTok).

You don't need to predict trends before they happen, just as they're happening, and before peak popularity. This is the safest (not that such a strategy is safe) way of getting into trends; not by trying to predict them, but by hitching a ride. Returns aren't the same (relative to being extremely early to a trend), but risk is equally lessened.

Investing Rules

Implementing firm investing guidelines aims to conquer irrationality and prevent bad decision-making, especially as the result of subjective viewpoints and emotional decisions. As follows:

- Nothing lasts forever
- No woulda, shoulda, coulda
- Don't be emotional
- Diversify
- Prices don't matter

Nothing Lasts Forever

Just in viewing this simple chart of bitcoin over the past five years (this writing taking place in 2022), it is easy to see that no market condition has lasted for any significant period of time. Periods of bull runs are typically followed by sharp downtrends, while historical market cycles of similar asset classes insinuate that most cryptocurrencies won't exist at scale within a few decades.[3] No market condition is permanent, no asset will thrive permanently, and the wider performance of nation-state economies are variable. This isn't stated to be a downer, but rather to impart the value of detaching from short-term FOMO and JOMO and the benefits to be derived from understanding both short-term and long-term cycles in assets and markets.[4]

---

[3] Most likely as the result of some filter event like the dot-com crash.
[4] Fear of Missing Out and Joy of Missing Out. On this note, I recommend Ray Dalio's *The Changing World Order* as a fantastic read on nation-state economic cycles.

No Woulda, Shoulda, Coulda

This rule is taken from the legendary stock trader and host of the show *Mad Money*, Jim Cramer. The idea is represented through no woulda, no shoulda, and no coulda. When investment don't play out as predicted (and plenty surely will), take a few minutes to think about how you can learn and improve. Doing so allows room for reflection and improvement while simultaneously maintaining sanity. Don't think about what you *would* have done, what you *should* have done, or what you *could* have done. Don't beat yourself up about losses, and don't let wins get to your head.

Don't Be Emotional

Emotion is the antithesis of analysis-based trading. Emotion, more often than not ("not" simply due to the random occurrence of making a good decision through a bad process), will only hurt you and take away from your investing strategy. Some people are naturally comfortable with the risk and emotional rollercoaster of investing, especially in volatile assets; if you're not, it's best to adjust your investment strategy to fit the personality.

While all that may seem a bit over the top, just wait until you enter into a risky position and try to go to sleep, or, worst of all, sell right before a massive pump. RIP.

Diversify

Diversification counters risk. While you both assume and are likely looking for a certain level of risk (due to risk and reward largely correlating) by investing in crypto, there is certainly a degree of risk you're not willing to assume. Diversification helps you stay within that maximum load of risk.

Generally, investors in the crypto space should keep a somewhat diversified portfolio, no matter how much belief may be held in a certain project. Fund allocation should (usually) be split between bitcoin, Ethereum or ETH alternatives (such as Cardano and Solana) and various altcoins, along with some cash. While exact percentages vary depending on individual situation (35/25/30/10, 60/25/10/5, 20/20/40/20, etc.), investing in a diversified manner across various aspects of the cryptocurrency market is a sustainable way to invest, capture gains across the market, and lower the impact of bad investments.

However, all that said, the crypto market is somewhat unprecedented. Some traders put most of their money into small-cap altcoins, while others dollar-cost average bitcoin and touch nothing else. At the end of the day, establish a strategy that fits your situation, resources, and personality, and then diversify within the boundaries of that strategy.

Prices Don't Matter

Given that supply and initial price can both be set, price itself is largely irrelevant without context. Just because Binance Coin is at $500, and Ripple is at $1.80 doesn't mean that BNB is worth 277x the value of XRP. Rather, the two coins are currently within 10% of each other's market cap.

When a cryptocurrency is first created, supply is set by the team behind the asset. The team may choose to create 1 trillion coins, or 10 million. So, looking back at XRP and BNB, it can be seen that Ripple has roughly 45 billion coins in circulation, and Binance Coin has 150 million. In this way, price doesn't really matter. A coin at $0.0003 can be worth more than a coin at $10,000 in terms of market cap, circulating supply, volume, users, or utility.

Price matters even less due to the advent of fractional shares, which lets investors invest any amount of money in a coin or token regardless of price. So, while price is still half of the market cap equation (price per unit x number of units = market cap), the second half of the equation can be set from the start. Plenty of other metrics should be considered before price, and absolute price should not factor into investment decisions.

Charts

Charts form the basis by which price can be examined and patterns can be found. Charts, on one level, are simple, and on another, deep and complex. To develop a thorough understand of charts, we'll begin with several different types. As follows:

- ♦ Line Chart
- ♦ Candlestick Chart
- ♦ Renko Chart
- ♦ Point & Figure Chart
- ♦ Heiken-Ashi Chart

Line Chart

A line chart is a chart that represents price through one single line. Most charts are line charts because, although they contain less information than popular alternatives, they are extremely easy to understand. Robinhood and Coinbase (both of which target their services towards less experienced investors) set line charts as the default chart type, while institutions aimed towards a more experienced audience, such as Charles Schwab and Binance, use other chart forms.

Candlestick Chart

Candlestick charts are a much more useful form of displaying information about a coin and are the chart of choice for most investors. Within a given period, candlestick charts have a wide "real body" and may be red or green (the other common color scheme being empty and filled real bodies). If it is red (filled in), the close was lower than the open (meaning it went down). If the real body is green (empty), the close was higher than the open (meaning it went up). Above and below the real bodies are the "wicks" also known as "shadows." Wicks show the high and low prices of the period's trading.

So, combining what we know, if the upper wick (aka the upper shadow) is close to the real body, the high the coin or token reached during the day is near the closing price, and the opposite also applies. You will need to have a solid understanding of candlestick charts, and services such as TradingView (tradingview.com) are a great way to get comfortable.

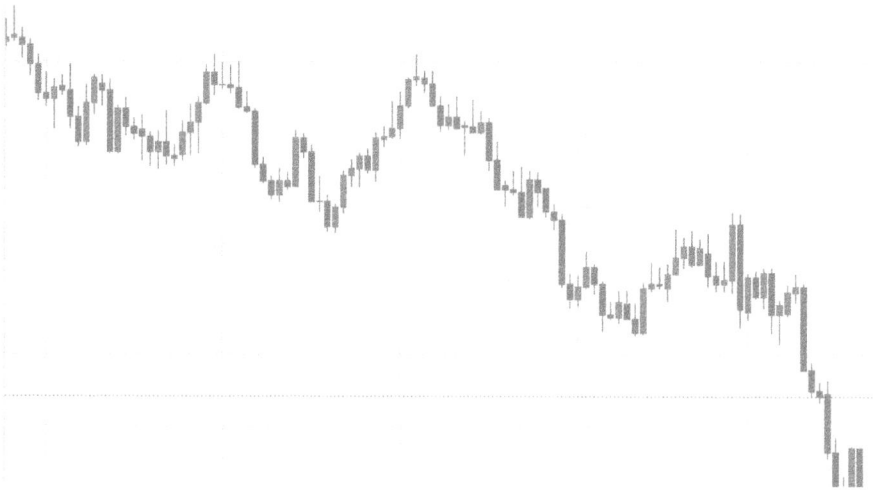

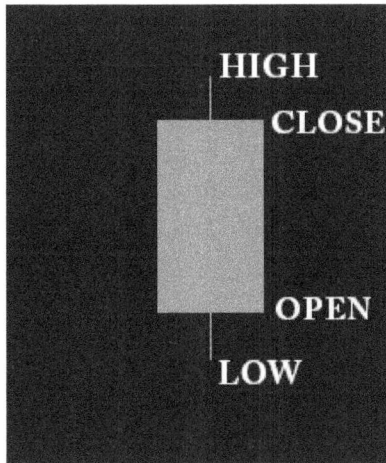

*Candles are usually either paired as red (down) and green (up) or filled (down) and empty (up). For example, this candlestick is green (grayscale aside), meaning it represents an upward price movement. Note that the "open" and "close" levels reverse in order on bearish candlesticks (red or filled) since such candles start lower than they opened, while the opposite is true for bullish (green or empty) candles.

Renko Chart

Renko charts only show price movement and ignore time and volume. Renko comes from the Japanese term "renga," meaning "bricks." Renko charts use bricks (represented as boxes), typically in the color scheme of red/green or white/black. Renko boxes only form at the top or bottom right corner of the proceeding box, and the next box can only form if the price passes the top or bottom of the previous box. For example, if the predefined amount is "$1" (think of this as similar to time intervals on candlestick charts), then the next box can only form once price passes either $1 above or $1 below the price of the previous box. These charts simplify and smooth out trends into easy-to-understand patterns, notably making support and resistance easier to discern.

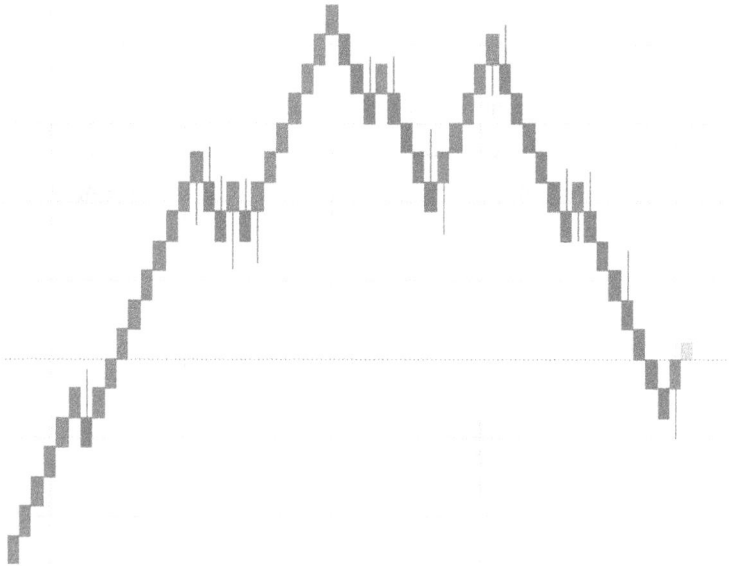

Point & Figure Chart

While the point and figure (P&F) chart isn't as well-known as others on this list, it does have a reputation as one of the simplest charts for identifying good entry and exit points. Like Renko charts, P&F charts don't directly account for the passage of time. Rather, X's and O's are stacked in columns, and each letter represents a chosen price movement (just like the blocks in Renko charts). X's represent a rising price, and O's represent a falling price. Consider this sequence:

```
    X
X O X
X O
X
```

Let's say the chosen price movement is $10. We start at the bottom left of the above visual: the 3 X's indicate that the price rose $30, the 2 O's signify a $20 drop, and the final 2 X's represent a $20 rise. Time is irrelevant, and trends are smoothed out.

Heikin-Ashi

Heikin-Ashi (hike-in-aw-she) charts are a simpler, smoother version of candlestick charts. They function almost the same way as candlestick charts (through candles, wicks, shadows, etc.), except Heikin-Ashi charts smooth price data over two periods instead of one. This essentially makes Heikin-Ashi preferable to many traders versus candlestick charts because patterns and trends can be more easily spotted and false signals (small, meaningless moves) are, in large part, omitted.

That said, the simpler appearance does obscure some data relative to candlesticks, which is part of the reason why Heikin-Ashis haven't yet replaced candlesticks. I suggest that you experiment with both chart types to figure out which best fits your style and ability to discern trends.

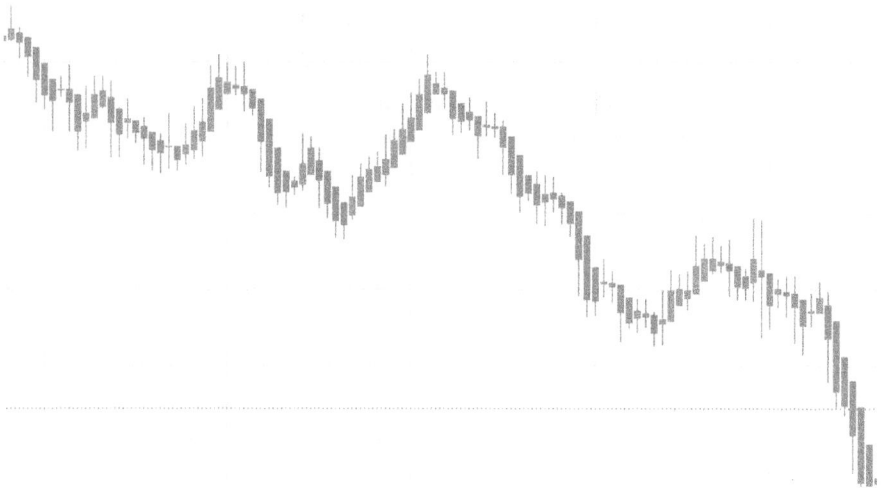

Note that the trends on the Heikin-Ashi chart are smoother and more discernible than on the candlestick chart on page 82.

Patterns

This section contains the essential chart patterns used in technical analysis. All are important, and modern charting tools allow for patterns such as these to be automatically, or at the very least easily, identified.

As follows:

- Triangle
- Rectangle
- Head and Shoulders
- Double/Triple Bottom/Top
- Doji
- Morning Star/Evening Star
- Abandoned Baby
- Two Black Gapping

Triangle

Triangle patterns can be either symmetrical, ascending, or descending. Ascending triangles consist of a horizontal trendline and a diagonally rising lower trendline, descending triangles consist of a horizontal lower trendline and a diagonally sinking upper trendline, and symmetrical triangles represent two trend lines and a shrinking price range. Ascending triangles signal bullish breakouts, descending triangles signal bearish breakouts, and symmetrical triangles signal potential breakouts in either direction.

(tradingview.com) Symmetrical Triangle #1

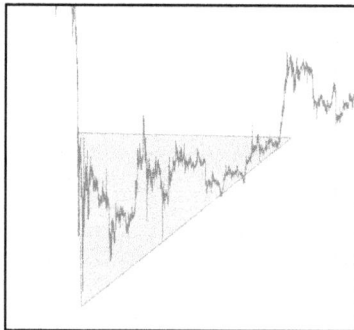

(tradingview.com) Ascending Triangle

Rectangle

Rectangle formations are continuation patterns signified by near-equal successive tops and bottoms. Rectangles have been found to be roughly 80% accurate and breakouts reliably extend as far as the trading range (the width) of the rectangle.

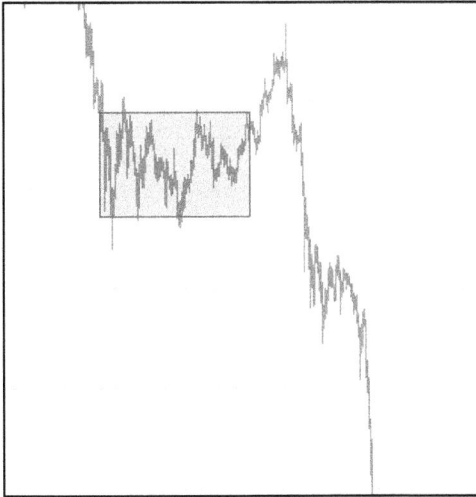

(tradingview.com) Rectangles #14

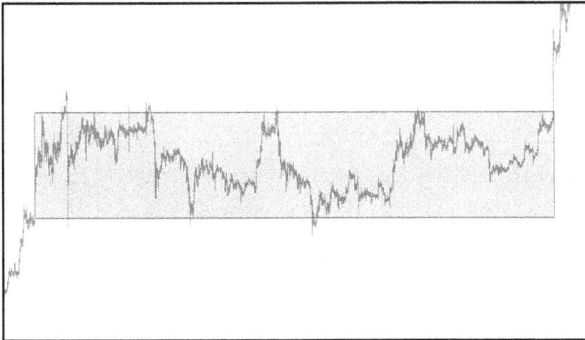

(tradingview.com) Rectangles #2

*Note that the resulting uptrends in both cases were roughly equivalent to the height of the respective rectangles.

Head and Shoulders

Head and shoulders are statistically the most accurate price action pattern, being correct roughly 85% of the time. The pattern consists of a baseline price and three peaks; the middle peak is called the "head" and is sandwiched between two "shoulders." The troughs of the shoulders form the "neckline" price. Head and shoulder formations indicate a bearish reversal, while inverse head and shoulder patterns are bullish.

(tradingview.com) Inverse Head and Shoulders #1

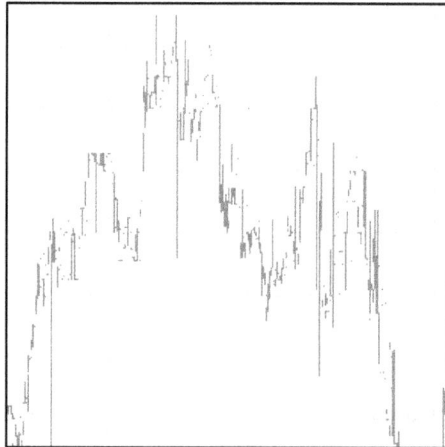

(tradingview.com) Head and Shoulders #2

Double/Triple Bottom/Top

Double tops, double bottoms, triple tops, and triple bottoms signify reversals. Each is signified by the corresponding number of distinct peaks or troughs. The formations, as a whole, are 75% to 80% accurate.

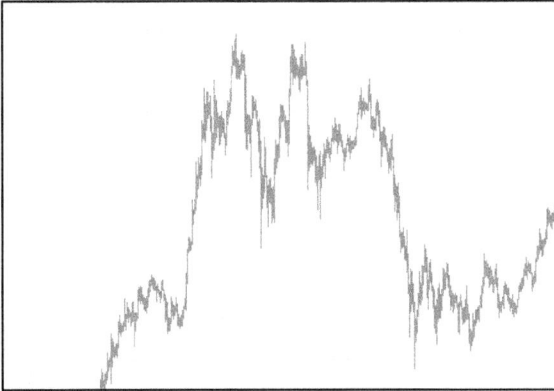

(tradingview.com) Double Top
(tradingview.com) Triple Bottom

Doji

Dojis are one-candle formations characterized by small trading ranges and long shadows. Standard dojis and long-legged dojis have shadows of equal length, dragonfly dojis have long lower shadows, gravestone dojis have long upper shadows, and four-price dojis are one thin, horizontal line with no shadows. Dojis often signify reversals but are much better used in conjunction with other bars to form stronger indicators.

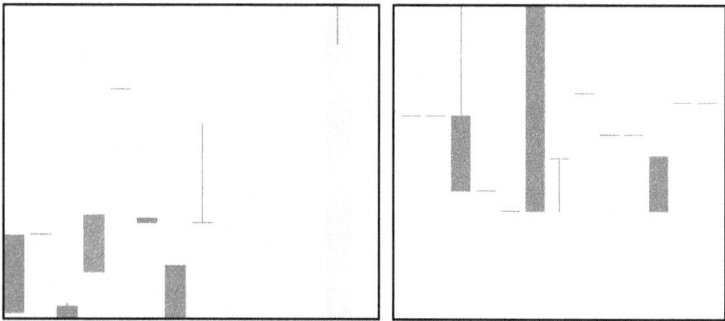

(tradingview.com) Gravestone Doji
(tradingview.com) Dragonfly Doji
*Middle red candle left, middle green candle right.

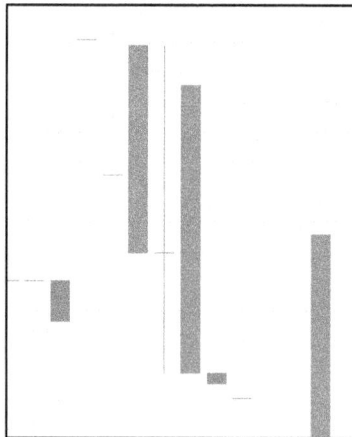

(tradingview.com) Long-Legged (Neutral) Doji
*Middle green candle.

Morning Star/Evening Star

The morning star and evening star are three-bar patterns, morning being bullish and evening bearish. The morning star constitutes a short-bodied candle between a long red candle (left) and a long green candle (right). The evening star constitutes a long green (left), a short-bodied middle candle, and a long red (right). Morning stars happen at the bottom of a downtrend, while evening stars happen at the top of uptrends. Both indicate reversals.

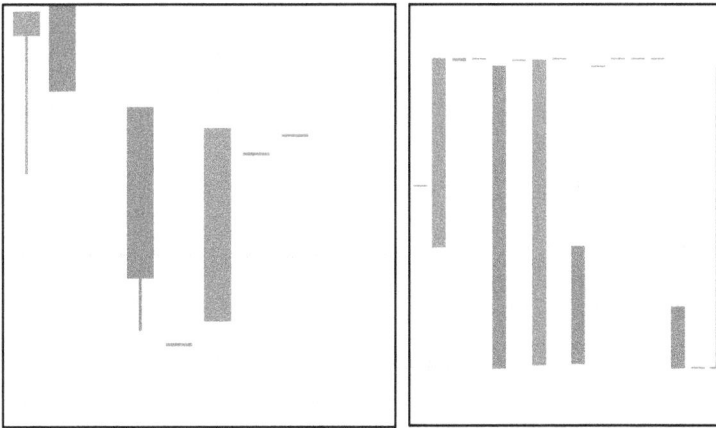

(tradingview.com) Morning Star
(tradingview.com) Morning Star #2

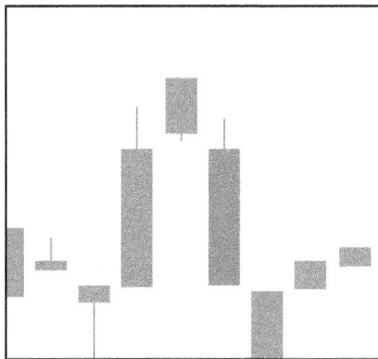

(tradingview.com) Evening Star

Abandoned Baby

An abandoned baby is a three-bar reversal pattern.[5] The bullish pattern constitutes a large candle, a doji that gaps lower, and a third large green candle that gaps higher. The bearish equivalent constitutes a large green candle followed by a doji that gaps upwards and, finally, a large-bodied red candle that gaps downward. This formation is known to be apt at predicting short-term reversals.

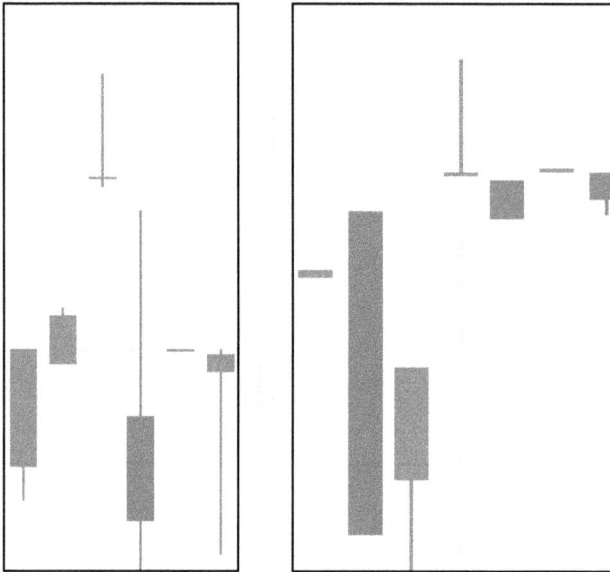

(tradingview.com) Abandoned Baby #1
(tradingview.com) Abandoned Baby #2

---

[5] Yes, I also wonder why that specific name was chosen.

Two Black Gapping

Two black gapping is a bearish continuation pattern which constitutes two red candles formed at the top of an uptrend, the second of which gaps below the first. This is indicative of a short-term trend remaining bearish. The formation is accurate roughly 70% of the time.

(tradingview.com) Two Black Gapping #1

(tradingview.com) Two Black Gapping #2

Indicators & Oscillators

This section covers a few select indicators and oscillators.[6]

As follows:

- Bollinger Bands
- Parabolic SAR
- Fibonacci
  - o Fibonacci Retracement
  - o Fibonacci Fans
  - o Fibonacci Arcs
- Moving Average
- RSI
- MACD
- Stochastic

---

[6] Oscillators, pronounced with a silent C, are defined as indicators displayed through a value that fluctuates between an upper and lower limit.

Bollinger Bands

Bollinger bands represent a class of indicators known as price envelopes. Price envelopes are bands that fluctuate around a price and identify range, support, and resistance. By identifying a range, traders can get an idea of whether prices are high or low on a relative basis.

Bollinger bands are used in conjunction with a moving average, which is a line called the "middle band" that mostly stays within the upper and lower band. The closer the price is to either the upper or lower band, the stronger the trend is. The closer the price is to the upper band, the more overbought the price, while the closer the price moves to the lower band, the more oversold the price.

Since approximately 90% of price action occurs between the two bands, the resulting 10% of price action is likely to be indicative of major breakouts. Stop losses can be placed at breakout prices once breakouts are confirmed.

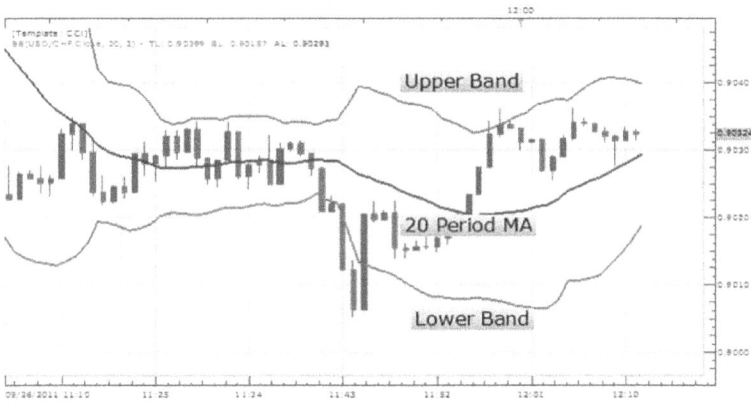

Bollinger Bands

Parabolic SAR

The Parabolic SAR (stop and reverse) is an indicator created by J. Welles Wilder[7] to determine trend direction and identify reversals. On a chart, this indicator appears as a series of dots. Dots below the price are bullish, while dots above the price are bearish. Dots that cross the price signal a trend reversal. Parabolic SARs work best during a strongly trending market, as opposed to a choppy or sideways-trading market. The Parabolic SAR is best used in combination with other indicators that aid in determining the strength of trends as opposed to the occurrence thereof.

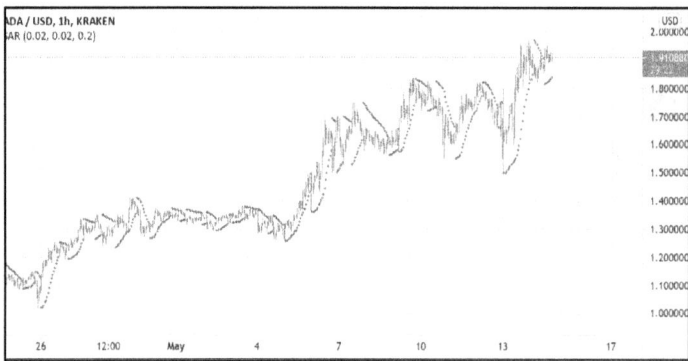

(tradingview.com) Parabolic SAR

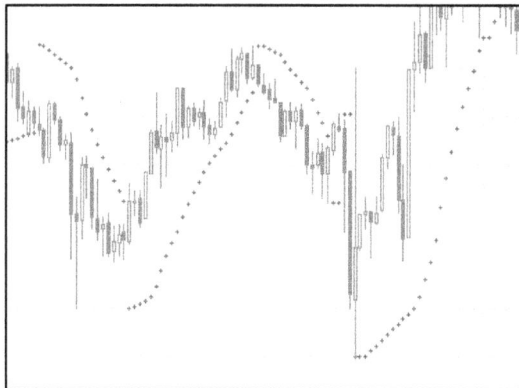

(tradingview.com) Parabolic SAR #2

---

[7] Wilder also notably created the RSI (relative strength index).

Fibonacci

You have likely heard of either the Fibonacci numbers, the golden spiral, or the Fibonacci sequence. Fibonacci numbers and tools fill a top spot in the indicator market and maintain a cult-like following. The Fibonacci sequence was invented by Leonardo of Pisa (1180–1250), an Italian who grew up in North Africa during the Middle Ages.[8] His nickname was Fibonacci.

He wrote a work called "Libre Abaci," which roughly translates to "The Book of Calculation." The book popularized the Hindu-Arabic numeral system, as opposed to the then-used Roman numerals. Within the book, the sequence of numbers that later became the Fibonacci sequence was used to calculate the growth of a rabbit population.[9] The question goes as this: how many pairs of rabbits will there be in one year, assuming that one initial pair of rabbits produces another pair of rabbits each month following a one-month infertility period? (Assuming each pair breeds indefinitely). The result is an equation that adds the sum of the two previous terms to get the next term:

$$F(n) = F(n) + F(n-1)$$

So, starting with 1 pair of rabbits, the following ensues:

$1 + 0 = 1$	$0 + 1 = 1$	$1 + 1 = 2$	$1 + 2 = 3$	$2 + 3 = 5$	$5 + 3 = 8$
$8 + 5 =$ 13	$8 + 13 =$ 21	$13 + 21 =$ 34	$21 + 34 =$ 55	$34 + 55 =$ 89	$55 + 89 =$ 144

---

[8] His name, as per various primary sources, may also be Leonardo Fibonacci, Leonardo Bonacci, or Leonardo Pisano.

[9] Fibonacci himself didn't regard his calculations as important. Instead, in 1877, the mathematician Edouard Lucas published studies involving the sequence which he labeled "the Fibonacci Sequence" in honor of the original author.

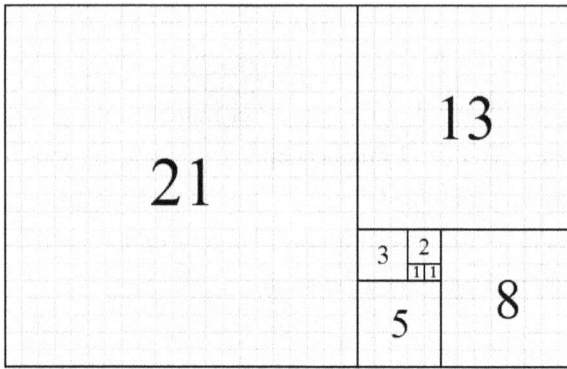

The Golden Numbers

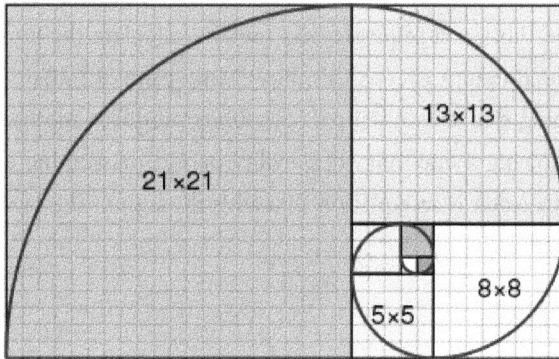

The Fibonacci Spiral

The resulting sequence and the equation (the summing of the previous two terms) is called the Fibonacci sequence. The golden spiral, in turn, is derived from the Fibonacci sequence. Both the Fibonacci spiral and the golden numbers involve the "golden ratio" of 1.618. The Fibonacci sequence and the golden ratio have been found all over the natural world and prove to be a naturally occurring pattern, being found in pinecones, flowers, various fruits and vegetables, honeybee colonies, and even the human body.

Fibonacci numbers have since been found to hold sway in the stock market. All Fibonacci market tools involve a trendline (often multiple) drawn between two points and primarily indicate support and resistance.

Fibonacci Retracement

The Fibonacci retracement indicator plots retracement lines as per the Fibonacci sequence. A retracement is a minor pullback or change in direction, so a retracement line is a line that indicates where support and resistance (hence, pullbacks and change in direction) are likely to occur.

Fibonacci retracements are created by drawing a trend line between two points—typically a low and a high, or vice versa. Six horizontal lines are then automatically drawn at points that intersect the original trend line. These interception points occur at Fibonacci levels of 0.0%, 23.6%, 38.2%, 50%, 61.8%, and 100% (in fraction form: 0, 0.236, 0.382, 0.5, 0.618, and 1). These lines identify possible areas of support and resistance.

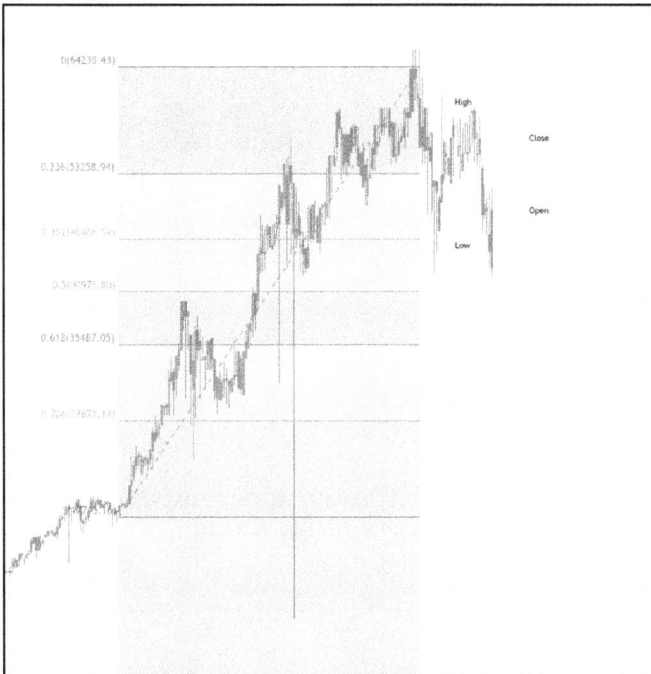

(tradingview.com) Fibonacci Retracement

Fibonacci Fans

Fibonacci fan lines are similar to the Fibonacci retracement indicator. First, a trend line is drawn between two points (typically an extreme point—either a high or a low). Then, four trend lines are drawn from the initial point and pass through an invisible vertical line below the second extreme point at the Fibonacci percentage levels described previously.

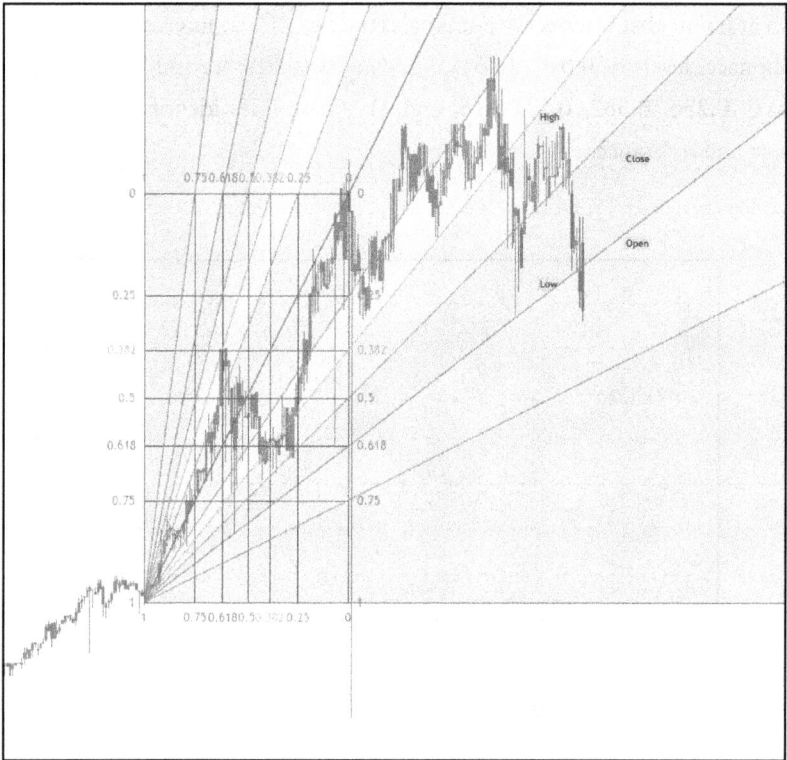

(tradingview.com) Fibonacci Fans

Fibonacci Arcs

Fibonacci arcs are half-circles that extend outward from a vertical line extending from the second of the two extreme points. The arcs of the half-circle are drawn at points that interest the trend line at Fibonacci levels.

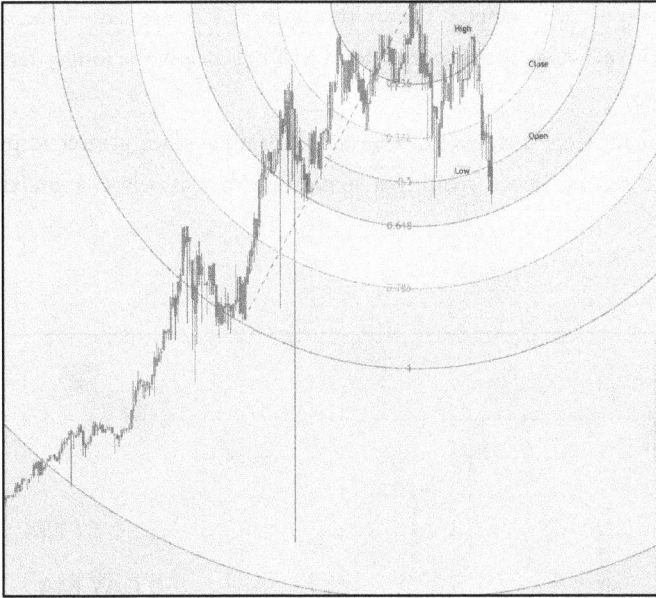

(tradingview.com) Fibonacci Arcs

Other Fibonacci Indicators:

♦ Fib Wedge - set of Fibonacci-based arcs.
♦ Fib Channel - parallel Fibonacci-based trendlines.
♦ Fib Circles - 11-layered Fibonacci-based circle.
♦ Fib Time Zones - vertical lines that represent potential price action based on Fibonacci-based time increments.
♦

Moving Average (MA)

Moving averages are lagging indicators that signal support, resistance, and momentum through a singular, smoothed-out line calculated in accordance with a time frame. Hence, one may say "5-day MA" or "100-day MA."

Often, moving averages are used in pairs; in such cases, crossovers signal a change in momentum (positive change if the shorter-term MA crosses above the longer-term MA or negative change if the shorter-term MA falls below the longer-term MA).

Exponential moving averages (EMAs) are MAs that place greater importance on recent price action, hence creating a signal line that associates more closely with price.

Moving Averages[10]

---

[10] Image credit to dailyfx.com

RSI

The RSI (relative strength index) is a momentum oscillator that measures the strength or weakness of price trends and, therefore, the likeness of reversals. RSIs trade within a range of 0 to 100: a value over 70 indicates an overbought condition (a sell), and a value under 30 indicates an oversold condition (a buy).

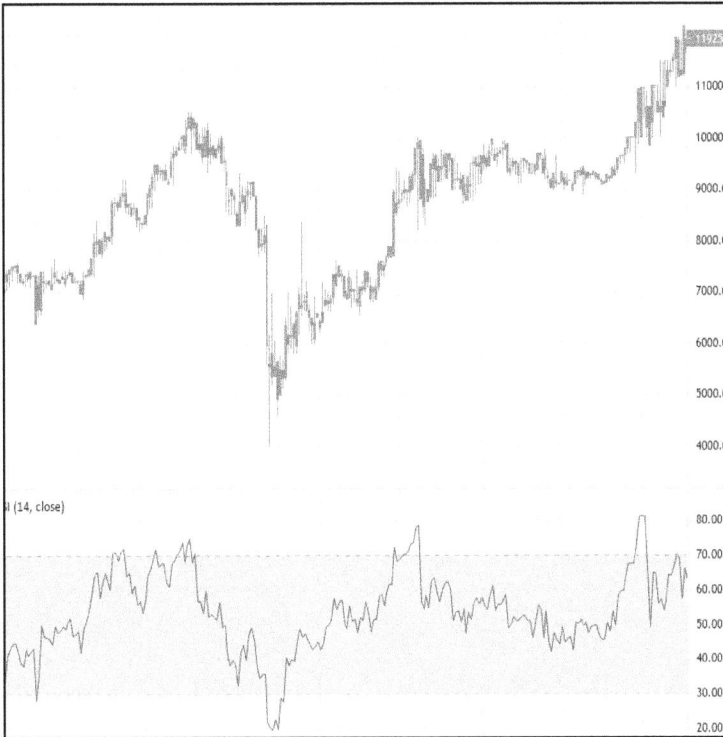

(tradingview.com) RSI Oscillator

MACD

The MACD is a momentum oscillator that identifies potential trend reversals through a change in momentum. The MACD operates through the MACD line (found by subtracting the 26-day EMA from the 12-day EMA), the signal line (a 6-day EMA), and a histogram, which plots divergence.

Crossovers between the two lines indicate a change in momentum; a bullish crossover being a MACD cross above the signal line and a bearish crossover being a MACD cross below the signal line. The height of the histogram is indicative of momentum strength.

(tradingview.com) MACD Oscillator

Stochastic

The Stochastic Oscillator is a leading momentum oscillator that aims to signal reversals and momentum changes before they happen. The stochastic value is plotted between 0 and 100; above 80 indicates overbought conditions and under 20 indicates oversold conditions. The Stochastic Oscillator is known to be quite accurate.

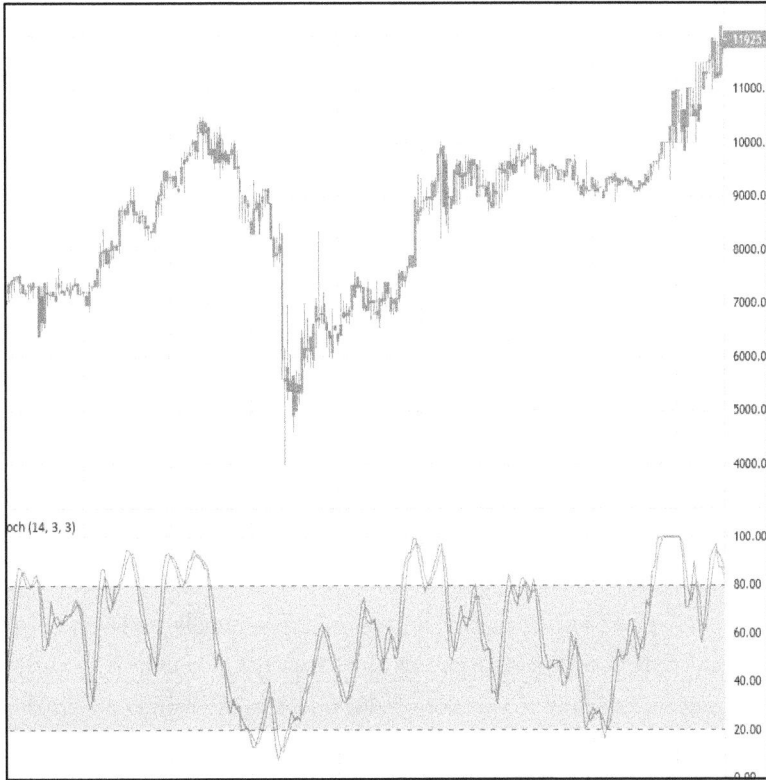

(tradingview.com) Stochastic Oscillator

Investing Strategy

This concludes a look into on-the-ground cryptocurrency trading types, patterns, charts, indicators, and oscillators. The last piece of the puzzle, namely, is fitting the pieces together, and this is done through an overarching investing strategy, beginning with infrastructure.

Investing infrastructure is the software that surrounds and powers an investor. The modern game is played online; and software helps you identify potential investments, evaluate those investments, and make trades. Setting up an investing framework begins with choosing an exchange, the most popular of which (in the US) are Coinbase, eToro, Binance US, and Kraken. The most popular global exchanges are (respectively) Binance, Coinbase Exchange, Huobi Global, Kraken, and Bitfinex.[11]

Then, I suggest setting up a means of performing chart analysis, such as through TradingView. Following setup on these platforms, other useful software includes sentiment analyzers (santiment.net), calendar websites (coinmarketcal.com) bot software, market data websites (onchainfx.com, coinmarketcap.com), airdrop websites (airdrops.io, airdropalert.com), and bitcoinvisuals.com. With these software tools and more, you'll have the toolbox within which your trading can flourish.

Once infrastructure is operational, the next step is to establish a set of rules that match the way you want to trade and the way you naturally trade best. Rules are necessary since the loss or gain of money, for most, is a very emotional experience. Harvard researchers found that making good trades in the market affects the brain in the same way as cocaine.[12] Beyond a base degree of natural emotions, money-related decisions are usually driven by some combination of family history, insecurity, fear, and greed, all accompanied by some degree of rationality.

---

[11] Based on current market data from coinmarketcap.com. Liable to change.
[12] (n.d.). Functional Imaging of Neural Responses to Expectancy ... - CiteSeerX.

So, two ideas must be understood: psychology affects how *you* trade, and psychology affects how other people, and hence *the entire market* moves. Bull markets and uptrends are based on greed, euphoria, optimism, and trust. Bear markets and downtrends are based on anxiety, denial, fear, and panic.

Despite all this, investing is really a numbers game, and excluding emotion from trading decisions is necessary in creating holistic success. Establishing trading rules, both online and off, is helpful in ensuring consistently profitable decision-making. Though many find these steps to be unnecessary, investing safeguards have proven time and time again to be anything but.

When To Stop

If making short term moves, stop trading after three consecutive wins or losses. This can be a painful rule, but is as effective as it is painful. Three consecutive good trades incur an emotional high, while three consecutive losing trades incur emotional lows. Either way, these situations lead to emotional states of mind and impair one's ability to think rationally and make good decisions. At the very least, take a long break, though the best thing to do is simply stop trading and resume at a later date. Of course, as with all these rules, adjust this to fit your personality; if you're experienced and consistently profitable, you might move to stopping at 4 trades. If you're just starting out or know yourself to be prone to emotional trading, maybe move it down to 2.

Mantra

Having a few choice words to repeat to yourself can be a quick way to reach a confident and focused state of mind. While it is much better for you to choose something that's truly meaningful to you (as the kids says, "hits different"), here are a few mantras for inspiration:

+ "I am calm, focused, and confident."
+ "I attract opportunity."

- ♦ "Slow and steady wins the race."
- ♦ "I am disciplined and patient."
- ♦ "Look at me now, mom."

Trade Size

Don't increase or decrease the money you're putting into a position because "you're feeling it." Adjust positions based on risk; nothing else. As a rule, it is best not to change a standard amount of money put into trades unless such a decision is based upon thought-out and comprehensive analysis. Have a conversation with yourself and ask why you want to make the investment you do. What's it founded on?

Trading Journal

Many investors keep a log of all trades; they'll write down entry and exit prices, the ticker, and any notes or observations. This isn't just an activity that benefits your trading; it is also fun and provides useful perspective throughout times of both up and down.

Intuition: A Note

Intuition may not fall directly into trading rules, but it does tie into risk mitigation, and investing rules are all about mitigating risk. Intuition derives from subconsciously realized patterns. An experienced, long-time investor might experience strong intuition based on patterns not consciously noticed. This, ultimately, is where the question of intuition must end. The more experienced you are, the better your intuition will be, while when just starting out, "I'm feeling it" will probably lead to a big, red loss.

Of course, the logical question is when have you have you become experienced enough; is it one year, three years, or so on? I say take steps to test your intuition over time while mitigating risk. So, perhaps enter a position with 1/10th of your normal volume, or perhaps don't trade, but track whether you would have been right and adjust from there. Just make sure to recognize intuition as intuition and not greed or fear in disguise, and don't use intuition as an excuse for a bad decision. And no matter how much you're feeling it, sticking to your investing plan is nearly always the better option.

Once infrastructure and rules are chosen and put into action\, the completion of the strategic investing trifecta requires the mastery (or at least a comfortable knowledge) of the down-in-the-dirt tools that inform trading decisions. This first involves a chart (candlestick, bar, Heiken-Ashi, Renko, etc.) followed by patterns (triangles, cup-and-handles, etc.) and indicators and oscillators (RSI, MFI, MACD, Bollinger Bands, Fib Arcs, MAs, etc.). Understanding economic data and ingesting market-wide research is just as important.

In combination, one should arrive at a structure with software to capture and assemble data and make trades, tools to figure out what those trades should be, and rules to maximize the effectiveness of such trades. Of course, no good strategy is set in stone; it must react and adjust based on new experiences and information, and you must pursue calculated and rational optimization among each area of strategy.

Strategy, ultimately, is the art of optimizing your experience in an environment before entering it and once in it, it's having a plan, using tools available to you, and working to improve such plans and tools based on conscious and sustained iteration. Doing this puts you ahead of most; and the disciplined pursuit of excellence, in any field as much as investing, will inevitability lead to long-term success.

Trading Terms

Bear Flag / Bull Flag

A bear flag is an indication on a coin or token chart that a downtrend is likely. A bull flag is the opposite of a bear flag and signals an uptrend.

Bear Trap / Bull Trap

A bear trap refers to false downtrend signal. This is referred to as a "trap" because traders who take a bear trap as an indication of a downtrend may short the coin or token; hence losing money when the price actually increases. A bull trap is the opposite of a bear trap.

Bear traps can sometimes be manipulated into existence. In such a case, a group of traders aims to quickly crash a cryptocurrency and then make a quick profit on the rebound. Those involved must own a sizable position of a given coin. Then, they all sell their positions at the same time, which bluffs the market into thinking a crash is occurring, which triggers more selling, causing stop losses to be hit and more selling to occur, leading to an even sharper fall. Those who set the trap proceed to buy back into their position at the lower price. Once the price rebounds, they make a profit.

Bear / Bearish / Bull / Bullish

To be a bear means that you think the price of a coin, token, or the value of the market as a whole is going to go down. Investors who think like this are considered "bearish" on the given asset. The opposite is to be bullish: a person who thinks a rise in value will occur is bullish on that asset.

Bubble

A bubble in investments refers to a time in which most of the market is going up, usually at an unsustainable rate. Often, bubbles will pop and trigger a large crash. For this reason, being in a bubble, whether referring to one coin or an entire market, is not a good thing.

Buy Wall

A buy wall occurs when a large limit order is placed to buy a cryptocurrency at a certain value. This "buy wall" can prevent the asset from falling below that value, since demand at that price far exceeds supply.

Confluence / Confluence Trading

Confluence occurs when multiple strategies and indicators are combined into one strategy. Confluence trading is an extension of this; it refers to a trader who utilizes confluence in their trading strategy.

Correction

A correction is a price movement downward after a quick jump or a peak in price. For example, a move from $10 to $25 may result in a correction to $20, at which price support is found.

Dead Cat Bounce

A dead cat bounce is a term that refers to a brief price recovery before a large crash.

Depth Chart

Depth charts graph buy and sell requests. The depth chart shows the crossover point at which transactions are quickly completed, which is the market price.

Dump

To dump, or dumping, refers to selling a large amount of cryptocurrency or to a large amount of a coin or token being sold. For example, "that coin is dumping," and "I'm dumping this coin."

Fill or Kill Order (FOK)

A fill or kill order is an order that must be executed upon immediately. If this doesn't happen, the trade will be canceled. FOKs are used to ensure that transactions involving large positions will be completed in a very short period.

Fundamental Analysis

Fundamental analysis is analysis of a coin or token through its fundamental metrics. Refer to the earlier section on fundamental analysis for a comprehensive overview.

Golden Cross

A golden cross is a chart pattern involving a short-term moving average (say, a 10-day MA) crossing above a long-term moving average (perhaps a 50-day MA). Golden crosses are bullish indicators.

Leverage

Investors can "leverage" their money by taking on debt. Say you have $1,000 and you take on 5x leverage; you're now able to invest $5,000 worth of funds. By that same function, 10x leverage is $10,000 and 100x is $100,000. Like margin trading, leverage allows amplified profits by, in a sense, renting money and pocketing the extra profit. However, leverage trading is very risky; unless you are an experienced trader and financial stable, leverage trading is not recommended.

Limit Order / Buy / Sell

When you mean to execute a trade, you may choose to have that trade executed in several different manners. One of such manners is through a market order, which executes orders immediately at the best market price available. The popular alternative is a limit order, which lets the buyer or seller choose the price at which they want to buy or sell.

For example, say a coin is trading at $200. If you choose to buy one coin with a market order, that order will immediately execute, perhaps at $200, or perhaps at $199 or $201.[13] If you place a limit order, you choose the price you want to purchase the one coin at. Maybe this coin is volatile, so you decide to place a limit buy order at $197 in the hopes of the price spiking down to that level at some point throughout the day and before recovering and continuing in an uptrend. In this case, the order

---

[13] This is called slippage. Slippage may be as high as 2-3%.

would only execute when the price of the coin hits $197 or below. Generally, limit orders are good for catching the price slightly below the market value at the time of the order, though limit orders can be set at any price for various other reasons. If you're placing and order and don't really care whether your purchase point is 2% lower or not (or whatever it may be), you can just set a market order and purchase the security instantly.

Long / Short (position)
Taking a long position means that an investor intends to hold an asset for the long-term; this generally means at least a few months. A short position is the opposite; the trader intends to get in and get out in a relatively short time period, whether minutes, hours, or days.

Margin Trading
Margin trading is a popular strategy whereby traders borrow funds to place trades. For example, someone with $10,000 may trade with 5x margin, giving them $50,000 of capital. If the trade works out, they repay the $50,000 (typically with interest or some type of fee) and keep the extra profit. Margin trading should only be performed by experienced investors—if trades go bad, many traders end up with more debt than money. So, the rewards are massive, but the risks are just as exceptional.

Market Capitalization (Market Cap)
The market cap of a coin is the total trading value. This can easily be calculated by multiplying the total supply of a coin by the price of the coin. For example, a cryptocurrency trading at $5 with a supply of 1 million units has a market cap of $5 million.

Market Momentum
The momentum of a market is the ability of that market to maintain periods of growth or shrinkage. A market that has been in the green for six months has strong momentum, while the same can be said if that market plunges into bear territory and stays in the red for significant periods of time.

Market Order

A market order is one several types of orders that can be placed to execute a trade. Market orders are immediately filled at the best market price available. The opposite, limit orders, allows the buyer to choose the price they want their trade to be fulfilled at. While market orders may result in the purchase point being slightly higher than a smart limit order, they allow for quicker entry.

Oversold / Overbought

An oversold cryptocurrency is experiencing much more selling pressure than buying pressure. As a result, it has been sold down to a price that is considered below its fundamental true value. Therefore, being oversold generally means that the security should rebound at least to its true value. Overbought is the opposite and occurs when a coin or token has been bought up to what may be considered an unjustifiably high price. Usually, if someone believes that a coin or token is oversold, they believe that it will go up, while if they believe it is overbought, they believe it will go down.

Pump

A pump is a rapid upward price movement in a coin or token.

Pump and Dump

A pump and dump is a scheme carried out by a large investor or, more typically, a group of large investors. In a pump and dump, the starting group will buy a significant amount of a coin or token. Other investors see the strong upwards pressure and buy in. Then, once the price has been significantly inflated, the original investors dump their shares and take profits. This practice is looked down upon since it is manipulative and causes most investors involved to lose money.

Resistance

Resistance is a price that an asset struggles to break through in an upwards manner. Sometimes, levels of resistance can be physiological. For example, bitcoin may hit resistance at $100,000, since many people place sell orders at the nice, round, and pleasant-sounding price of $100,000. When a resistance level is broken through, price can quickly climb. In a situation where bitcoin breaks past $100,000 following a period of strong resistance, the price may quickly climb to $105,000. Support is the opposite of resistance.

Sell Wall

A sell wall is a very large sell order at a limit price. Sell walls drive prices downwards. The opposite is a buy wall, which can stop a coin or token from falling beneath a certain price.

Slippage

Slippage can occur when a trade is placed through a market order. Market orders try to execute at the best possible price, but sometimes a notable difference occurs between the expected price and actual price. For example, say you want to buy 20 BNB for $1000, but you place a market order and only end up getting 9 BNB for $1000. Slippage is rarely this drastic, but regularly varies in the 1-3% range. When placing large orders, it is usually better to place a limit order as opposed to a market order. This eliminates the danger of slippage.

Support

Support is the price at which an asset struggles to break through in a downward manner since many investors are willing to buy the asset at that price and therefore buying pressure far exceeds selling pressure. Often, if a coin hits support levels, it will reverse into an uptrend. Support level bounces are often good short-term buy opportunities, though if support levels are broken through in a downward manner, a sharp and extended fall is likely.

Swing / Swing Traders

A swing is a dramatic reversal in price. Swing traders try to catch and trade upon market and asset-specific swings.

Tank / Tanked

Refers to a price taking a massive dive, e.g., "it tanked from $20 to $10."

Technical Analysis

Technical analysis looks at technical indicators to predict price movement. Technical analysts use historical data from charts to make their predictions. Refer to the technical analysis section earlier in the book for an extended look into technical analysis and various charting strategies.

Volatility

Volatility is the size of price change in a coin or token, and therefore the ability and likelihood that a price will change rapidly, whether in a positive or negative direction. So, a coin that moves 10% up one day, 27% down the next day, and 22% up the third day is more volatile than a coin that moves up 2%, down 0.5%, and up another 1%. Some coins, specifically stablecoins, have very little volatility, while other coins or tokens, typically those with a relatively small market cap, are extremely volatile and move up and down rapidly.

Wicks / Whiskers / Shadows

Whiskers are the lines extending from the colored bars on candlestick charts and refer to the low-high range of the given asset. Wicks, whiskers, and shadows are synonymous.

Informal Language

Whether in crypto chat rooms, watching crypto videos, or reading crypto news, you will certainly frequent the slang that dictates cryptocurrency culture. The culture, and informal language with it, is always evolving, but this section will get you started.

Bag

A bag refers to one's position. For example, if you own a sizable quantity in a coin, you own a bag of them.

Bag Holder

A bag holder is a trader who has a position in a worthless coin. Bag holders often hold out hope on worthless positions.

Dolphin

Crypto holders are classified through size of holdings. Those with extremely large holdings are called whales, while those with moderately sized holdings are called dolphins.

Flippening / Flappening

The "flippening" is used to describe the hypothetical moment when, if at all, Etherium (ETH) passed bitcoin (BTC) in market cap. The "flappening" was the moment when Litecoin (LTC) passed Bitcoin Cash (BCH) in market cap. The flappening happened in 2018, while the flippening has yet to occur, and, based purely on market cap, is presently unlikely.

Maker

Not to be confused with the Maker coin. A user becomes a maker when they place an order, and that order is not immediately fulfilled. That order becomes open and is placed in the order book until fulfilled.

Moon / To the Moon

Phrases such as "to the moon" and "it's going to the moon" refer to a cryptocurrency going up in value, typically by an extreme amount.

Vaporware

Vaporware is a coin or token that has been hyped up but has little intrinsic value and is likely to decrease in value.

Vladimir Club

A term that describes someone who has acquired 1% of 1% (0.01%) of the maximum supply of a cryptocurrency.

Weak Hands

Traders you have "weak hands" lack the confidence to hold their assets in the face of volatility and often trade on emotion, as opposed to sticking to their trading plan.

Whale

A whale is a person or entity that holds an extremely large position in a coin.

FOMO

An acronym meaning "fear of missing out."

FUD

An acronym meaning "fear, uncertainty and doubt."

HODL

A misspelling of "hold" that came about when a drunk bitcoin holder accidently typed "hodl" into a Bitcoin forum. Hodl is now synonymous with hold, though it is somewhat more cultured, and the word sometimes refers to an acronym meaning "hold on for dear life" though this is not the original meaning.

REKT

Rekt is slang for "wrecked" and is used to describe a bad trade or large loss.

Above & Beyond

Above & beyond constitutes just over forty terms that aren't essential knowledge but are still applicable and most certainly useful to know. Feel free to skip this section as per your interest level.

Arbitrage
Multiple exchanges trade the same cryptocurrencies at any given time and sometimes small differences in price occur between these exchanges. Arbitrate traders try to profit on this small margin by buying and selling the difference.

Bitcoin ATM
A Bitcoin ATM is an ATM from which bitcoin, and perhaps other cryptocurrencies, can be bought or sold. Bitcoin ATMs are gaining in popularity in parallel to the growth of Bitcoin and the wider cryptocurrency industry.

Bit
A microbitcoin, or bit, refers to one-millionth of a bitcoin.

Block Explorer
Block explorers are online services that track blockchain transactions and provide access to blockchain public ledgers. Etherscan is a popular block explorer.

Corporate Cryptocurrency
A corporate cryptocurrency is a cryptocurrency created or managed by a corporate entity.

Derivative
A cryptocurrency derivative is a financial product that derives value from an underlying asset, such as a coin or token, and allows investors to hedge their bets and mitigate loss.

Double Spend

Double spending occurs when a cryptocurrency is sent to two different wallets at the same time. Blockchains leverage various consensus models to stop double spending and such attacks are practically impossible to carry out on most major cryptocurrencies.

Dust Transaction

A dust transaction is an extremely small transaction. Sometimes, attackers flood networks with dust transactions to track down the transactional activity of certain wallets. Through a combined analysis of address activity, the attackers may be able to deanonymize the person or company behind the wallet.

Efficient Market Hypothesis (EMH) / Adaptive Market Hypothesis (AMH)

The EMH is an economic theory stating that the price of publicly tradeable assets reflects all publicly available information. Given the rise of coin and tokens that move in value simply due to hype, pump-and-dump schemes, and all the other ways in which an asset may uptrend without underlying fundamental reasons, the Efficient Market Hypothesis is not reflective of the current cryptocurrency market.

The Adaptive Market Hypothesis, AMH, proposes that financial markets are governed by the laws of biology. Those laws notably including the following: people act in their own self-interest, people make mistakes, and people adapt and select based on past actions. The AMH is more present in current crypto markets than the EMH.

Faucet

A faucet is a website that offers free cryptocurrency in exchange for information, such as an IP address. Faucets are scams and differ from Airdrops. Never give personal information away online, especially in return for promised "free" cryptocurrency.

Frictionless

A frictionless market is an ideal trading environment in which no costs or restraints exist on transactions. Frictionless markets are only theoretical.

Futures Contract

A futures contract is a pre-approved contract between two entities to fulfill a transaction at a certain price or date.

GitHub

GitHub is a collaboration platform for coders and software developers, in which open-source code can be shared, implemented, and improved.

Hard Cap

A hard cap is the maximum amount of money the creator(s) of a coin can raise during an ICO (initial coin offering).

Hexadecimal (Hex)

The hexadecimal system is a numbering system used to reduce the amount of work needed by computers. The Hex system is based off 16 symbols—0 through 9 and A through F.

Iceberg Order

An iceberg order executes a large order through many smaller orders. This allows the orders and the buyer or seller to stay somewhat discreet, as iceberg orders are used by those wanting to keep large transactions under the radar.

Instamine

An instamine occurs when coins are created as a single large batch, as opposed to slowly creating new coins through mining. Instamines are not very common and more susceptible to fraud.

Locktime

Some transactions come with a rule that delays the time at which the transaction can be validated and confirmed. This is called locktime. The vast majority of blockchains do not involve locktimes.

Mnemonic Phrase

A mnemonic phrase is synonymous to a seed phrase. Both terms describe 12-to-24-word sequences that identify and represent a wallet. It is like a backup password, and anyone with it has complete access to a cryptocurrency wallet. Make sure to securely store your mnemonic phrases.

Pegged Currency

A pegged currency, like a stablecoin, is a currency designed to remain at the same price of a designated asset, typically a bank-issued currency. USDT and DAI are two popular stablecoins pegged to the US dollar, meaning that 1 DAI and 1 USDT will be equivalent in perpetuity to the US dollar.

Proof of Authority (PoA)

A PoA algorithm gives a small number of users the authority to validate transactions. This saves computational power and creates a streamlined validation process.

Proof of Work (PoW)

A PoW algorithm is used to confirm transactions and create new blocks on a given blockchain. "Proof of work" literally means that work, through mathematical computation, is required to create blocks. The people who own the computers that perform the computations are miners.

Proof of Burn (PoB)

Proof of burn is a method of running a cryptocurrency by "burning," or destroying, a small percentage of the supply. Burning is the opposite of inflation, and the goal of burning is to increase the value of all other coins in the network by increasing scarcity. An extended breakdown of deflationary cryptocurrencies, inflationary cryptocurrencies, and burning is further back in the supply mechanisms section.

Race Attack

A race attack, which is a subset of double-spend attacks, occurs when two transactions are created with the same funds at the same time, with the goal of spending the money twice and doubling the initial amount.

Refund Address

A refund address is a wallet address that serves as a backup in case a transaction fails. If such an event occurs, a chargeback is given to the specified refund address. Refund addresses are not common, and in most cases, transactions sent to the wrong address are non-returnable.

Scrypt

A scrypt is an algorithm that encrypts data (notably keys) in a way that makes brute forcing the information very difficult. Scrypt-encrypted keys requires an enormous amount of computational power to hack (technically, to hash), which raises the barrier of entry for potential attackers.

Segregated Witness (SegWit)

A SegWit process allows more transactions to fit within one block by separating transaction signatures.

Selfish Mining

Selfish mining is a group mining strategy in which the miners strategically withhold blocks to increase profit.

Sharding

Sharding is the process of splitting up a blockchain network into smaller portions. In this way, transactions can be recorded and stored in only one shard, as opposed to going through every single node in the network. Sharding enables blockchain scalability and speed.

Solidity

Solidity is a programming language, just like Python or Java, which is used to write and develop Ethereum smart contracts.

Stale Block / Orphan Block

Due to the geographical separation of nodes in a blockchain network, multiple chains, each with multiple blocks, can simultaneously exist. Sometimes, two miners, each in different geographical locations, may hash (solve) a block at practically the same time. Both blocks are valid, and both reach different chains. One block will be "chosen" as the truth, with the choice being decided by the length of the chain. The block that did not get accepted is known as a stale or orphan block.

Tokenless Ledger
A tokenless ledger is a distributed ledger that does not require a cryptocurrency to function.

Token Lockup
Token lockups occur when a token or coin is not allowed to be transferred or traded.

Token Sale
A token sale, also commonly referred to as an ICO (initial coin offering), is a limited period of sale in which a set number of new coins are put up for sale in exchange for another cryptocurrency.

Transaction ID (TXID)
TXIDs are transaction identifiers used to reference transactions on a blockchain.

Turing Complete
Turing complete is a term that describes theoretical machines which can solve any computational problem, if given enough time, memory, and proper instruction.

Unspent Transaction Output (UTXO)
The UTXO is the amount of a digital currency left over once a transaction has been executed. A crypto wallet contains many UXTOs, all of which represent a certain amount of a coin or token. UXTOs can basically be thought of as smaller bills when paying with larger bills; they allow you to buy and sell fractions of coins, as opposed to whole quantities of such coins. In this way, UXTOs let you buy one-tenth of a bitcoin as opposed to one.

Virtual Machine

A virtual machine is a computer resource which allows users to run operating systems on computers. These operating systems act as separate computers despite existing purely virtually.

Wei

A wei is the smallest possible denomination of an Ether token.

Zero-Confirmation Transaction

A zero-confirmation transaction is a transaction that has not yet been confirmed on a blockchain.

Zero-Knowledge Proofs

Zero-knowledge proofs verify transactions without revealing information about the transaction, hence keeping the translator's information private while maintaining secure and efficient transaction mediums.

Acronyms

Considered in this section are the essential acronyms used in the crypto world. Many of the acronyms listed have a supporting definition elsewhere in the dictionary; if this is the case, search the index at the end of the book to identify expanded resources.

AMH

Acronym meaning "adaptive market hypothesis."

ATH / ATL

Acronym meaning "all-time high." This is the highest price a cryptocurrency reaches within a chosen time period. ATL means "all-time low" and is the lowest price a cryptocurrency reaches within a chosen time period.

BTD

Acronym meaning "buy the dip." May also be represented, with some salty language, as BTFD.

CEX

Acronym meaning "centralized exchange." Centralized exchanges are owned by a company that manages transactions. Coinbase is a popular CEX.

DYOR

Acronym meaning "do your own research."

EMH

Acronym meaning "efficient market hypothesis."

FA

Acronym meaning "fundamental analysis."

FOMO

Acronym meaning "fear of missing out."

FUD

Acronym meaning "fear, uncertainty and doubt."

ICO

Acronym meaning "initial coin offering."

JOMO

Acronym meaning "joy of missing out."

MCAP

Acronym meaning "market cap."

OCO

Acronym referring to the order type "one cancels the other." An OCO order lets two orders be placed at the same time. As soon as one of the orders becomes partially or fully filled, the remaining order will be automatically canceled. Popular exchanges such as Binance allow for OCO orders.

P2P

Acronym meaning "peer to peer."

PND

Acronym meaning "pump and dump."

ROI

Acronym meaning "return on investment."

SATS

SATS is shorthand for Satoshi Nakamoto, which is the pseudonym used by the creator of Bitcoin. A SATS is the smallest allowed unit of bitcoin, which is 0.00000001 BTC. The smallest unit of bitcoin is also referred to as a Satoshi.

TA

Acronym meaning "technical analysis."

TLT

Acronym meaning "thinking long term."

Acronyms: Above and Beyond

AML
Acronym for "anti-money laundering."

ASIC
Acronym for "application specific integrated circuit."

BFA
Acronym for "brute force attack."

DAO
Acronym for "decentralized autonomous organization."

DLT
Acronym for "distributed ledger technology." A distributed ledger is stored in many different locations so that transactions can be validated by multiple parties. Blockchain networks use distributed ledgers.

DPOS
Acronym meaning "delegated proof of stake." DPOS is a consensus algorithm designed to secure blockchains by using an election process between nodes.

ELI5
Acronym meaning "explain it like I'm 5."

IPO
Acronym meaning "initial public offering."

IEO
Acronym meaning "initial exchange offering."

MSB

Acronym meaning "money services business." MSBs transmit or convert money in some way, shape, or form. Cryptocurrencies are MSBs.

RSI

Acronym meaning "relative strength index." The RSI is a popular indicator used in technical analysis.

TOR

Acronym meaning "terms of reference."

TPS

Acronym meaning "transactions per second."

UTXO

Acronym meaning "unspent transaction output."

Visual Catalog

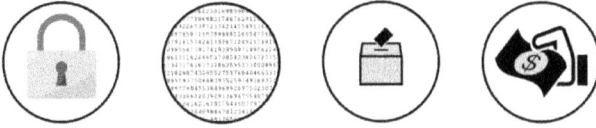

[1.0] The four token types – security, utility, governance, transactional.

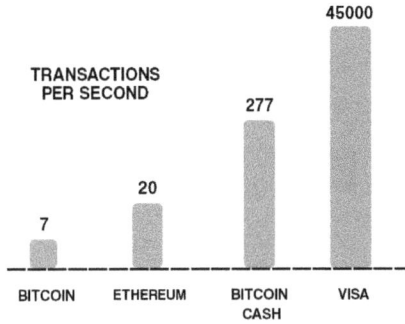

TRANSACTIONS PER SECOND

7 — BITCOIN
20 — ETHEREUM
277 — BITCOIN CASH
45000 — VISA

[1.1] Capacity transactions per second for Bitcoin, Ethereum, Bitcoin Cash, and Visa.

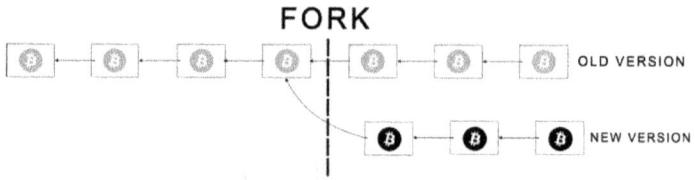

FORK

OLD VERSION

NEW VERSION

1.2 Visualization of a fork, in which a new version of a blockchain splits off from the old.

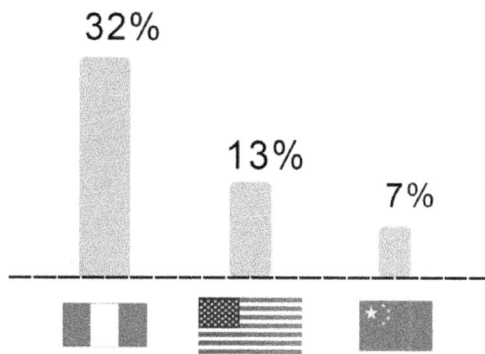

% OF RESPONDENTS WHO OWN CRYPTO

32%

13%

7%

1.3 Polled percent of respondents who own cryptocurrencies – Nigeria fills the top spot, and China, despite the many bans, remains around 7%.

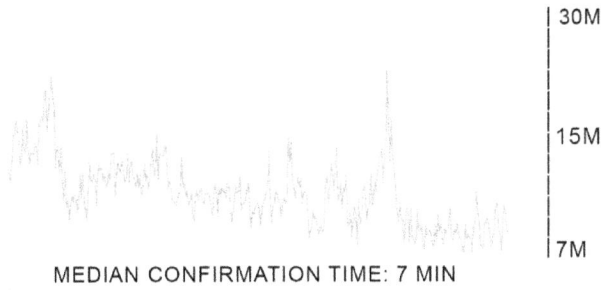

|30M

|15M

|7M

MEDIAN CONFIRMATION TIME: 7 MIN

[1.4] Median transaction confirmation time of the Blockchain network. (Assume variability.)

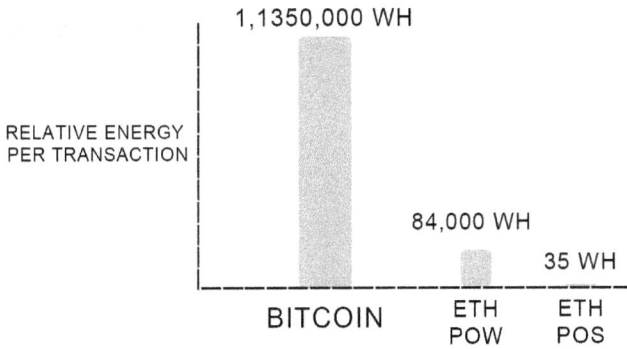

1,1350,000 WH

RELATIVE ENERGY
PER TRANSACTION

84,000 WH

35 WH

BITCOIN   ETH   ETH
            POW   POS

[1.5] Relative energy consumption per transaction of Bitcoin, Ethereum on its current proof-of-work system, and Ethereum 2.0 once a switch to proof-of-stake takes place.

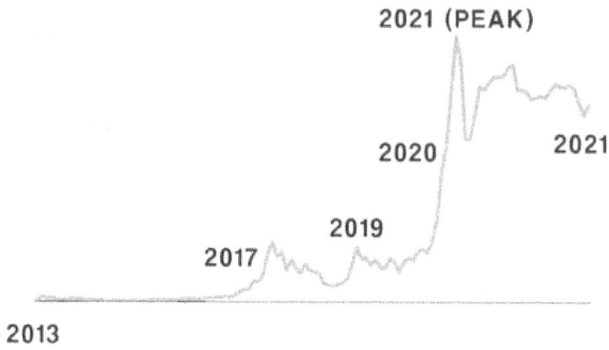

1.6 Simplified view of Bitcoin price action since inception. As of late 2021, the price has since returned to previous peaks.

1.7 Visualization of how a pair of public and private keys are used to establish a secure means of transacting.

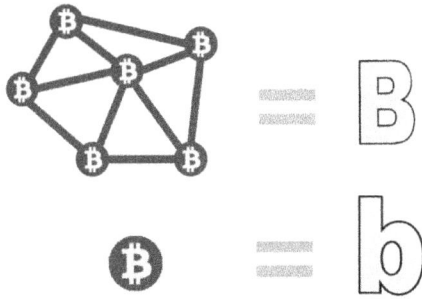

1.8 Bitcoin is capitalized when referring to the network, but not the unit. One may say "I bought 10 bitcoins on the Bitcoin network."

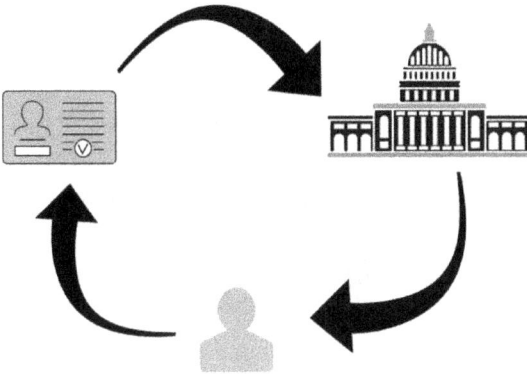

1.9 Most cryptocurrencies enable complete anonymity, but governments tend to require cryptocurrencies to be traded through centralized exchanges, which in turn require KYC laws (know-your-customer).

[2.0] Cryptocurrencies work through nodes, which are computers that join and assist in running the network.

[2.1] Cryptocurrency miners lend computational power to a crypto network. In return for providing such power, miners are given rewards in a given coin or token.

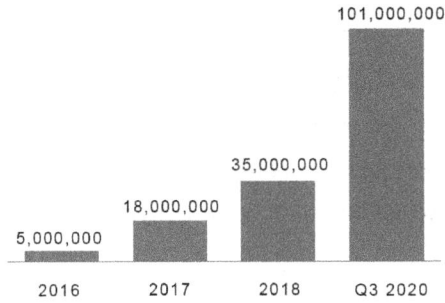

2.2 Number of identity-verified cryptocurency users from 2016 to Q3 2020. (Data from statista.com).

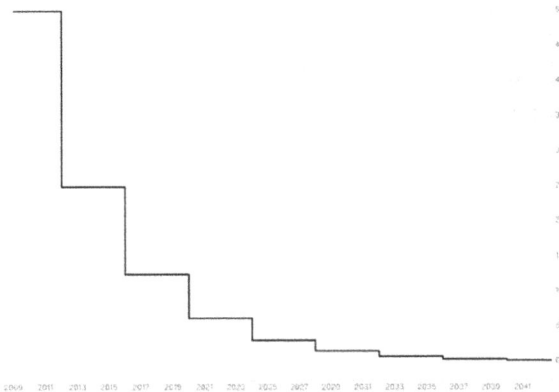

2.3 The number of coins added to the Bitcoin network every minute, which reduces as per four-year halvings. The most recent halving, in 2020, cut the block reward from 12.5 bitcoin per block to 6.25. The next will happen in 2024.

Key Players

The following is by no means a complete list, and merely serves to introduce you to notable players in the cryptocurrency space. 25 names are included, with special sections dedicated to Satoshi Nakamoto and Vitalik Buterin, who respectively created the first and second largest cryptocurrencies.

Satoshi Nakamoto

Satoshi Nakamoto is the individual, or possibly the group of individuals, who created Bitcoin. Not much is known about this mysterious figure, and his anonymity has spawned countless conspiracy theories. While Nakamoto has listed himself as a 45-year-old male from Japan on an official peer-to-peer foundations website, he uses British idioms in his emails. Additionally, the timestamps of his work align better with someone based in the US or the UK. Most believe that his disappearance was planned, and as Nakamoto currently holds a fortune worth more than $50 billion (equivalent to 1.1 million bitcoins) this anonymous billionaire, the father of cryptocurrency, could be the richest person in the world if Bitcoin goes up another few hundred percent.

Vitalik Buterin

The concept of the Ethereum platform was first put forward by Vitalik Buterin in 2013. Buterin is a Russian-Canadian programmer who through 2014 worked with a group of individuals to form what is today known as the Ethereum Foundation. Development was crowdfunded, with investors purchasing Ether (ETH) with Bitcoin (BTC) in order to get the project off the ground. Buterin's involvement with cryptocurrency began in 2011, when he started writing for *Bitcoin Weekly*, and later for *Bitcoin Magazine*, of which he was a co-founder. His involvement in the Bitcoin community helped to ground him firmly in the concept of cryptocurrency and move him toward what he viewed as a once in a generation opportunity.

While Ethereum is a public and transparent organization, Buterin's involvement with the project has given him something of a figurehead status, and his opinions and activities hold some sway over the financial performance of the network and underlying tokens.

Since Ethereum's inception, the Ethereum Foundation has driven the technical development of the network. At the beginning, Ethereum was functionally similar to other cryptocurrencies. However, with the foundation's process of adding new capabilities to the network, Vitalik Buterin and the team have brought forward now-ubiquitous features such as token sidechains, sharding, NFT capabilities, and more.

Charlie Lee
Founder of Litecoin.

Brian Armstrong
Founder of Coinbase.

Andreas M. Antonopoulos
Author of popular books such as *Mastering Bitcoin* and *The Internet of Money*.

Nick Szabo
Szabo conceptualized smart contracts for the first time in 1994 and designed Bit Gold, a decentralized predecessor to Bitcoin. Refer to the upcoming for more information on Bit Gold.

Brad Garlinghouse
CEO of Ripple.

Naval Ravikant
Ravikant is a well-known investor (book suggestion: *The Almanack of Naval Ravikant*) and advocate of cryptocurrency.

Cameron and Tyler Winklevoss

The Winklevoss twins are subjects of the movie "The Social Network" and former Olympic rowers. They made billions by investing early in Bitcoin and later launched the crypto exchange Gemini.

Michael Saylor

Saylor is the CEO of software firm MicroStrategy, which was one of the first major corporations to actively buy Bitcoin. The firm made billions off purchasing 114k bitcoins at a cost basis of $27k.

Matthew Roszak

Roszak is a venture capitalist and early adopter of Bitcoin.

Tim Draper

Draper is a venture capitalist who bought the bitcoins confiscated by US Marshals after taking down the Silk Road—the $18.7 million he paid for 30,000 bitcoin is now worth $2 billion.

Changpeng Zhao

Zhao is the founder of crypto exchange Binance.

Barry Silbety

Silbety founded the Digital Currency Group (DCG), a crypto conglomerate that manages entities such as CoinDesk and Grayscale.

Chris Larsen

Cofounder and chairman of Ripple.

Jed McCaleb

Cofounder of Ripple and founder of Stellar.

Haydeb Adams

Founder of Uniswap.

Andrew Cronje
Developer and founder of Yearn.finance

Kris Marszaler
Cofounder of crypto.com

Jesse Powell
Cofounder and CEO of Krane.

Gavin Wood
Cofounder of Polkadot.

Jihan Wu
Cofounder and CEO of Bitmain Technologies.

Charles Hoskinson
Cofounder of Input Output HK.

Gavin Wood
Founder and President at Web3 Foundation.

Sergey Nazarov
Cofounder of ChainLink and CEO of SmartContract.

History & Timeline

Constituting the following is a historical timeline of cryptocurrency, blockchain, decentralized technologies, and related matters.

1979

- Computer scientist Ralph Merkle describes an approach to public key distribution and digital signatures called "tree authentication" in his Stanford University Ph.D. thesis.

1982

- David Chaum describes a vault system for trustless online interaction in his 1982 UC Berkeley Ph.D. dissertation.

1983

- David Chaum develops the eCash platform, which launches with the goal of allowing people to transfer money anonymously over the internet.

1989

- Chaum goes on to found DigiCash, which bases itself upon the eCash concept and, like eCash, fails.

1991

- Stuart Haber and W. Scott Stornetta describe a cryptographically secured chain of blocks for the first time, as well as a process to timestamp digital documents.

1992

- Haber and Stornetta update their designs to incorporate Merkle trees in order to allow multiple document certificates to exist on a single block.

1997

- Adam Back introduces Hashcash, a PoW algorithm that aims to limit email spam through denial-of-service countermeasures.

1998

- Cryptocurrency is coined (no pun intended) as a term, the same year DigiCash goes bankrupt.
- Computer scientist Wei Dai launches B-money, which aims to be an anonymous, distributed electronic cash system.

- Nick Szabo starts work on a decentralized digital currency called Bit Gold, which never gets implemented.

1999

- Napster incurs a rise in the usage and popularity of peer-to-peer networks.

2000

- Stefan Konst publishes a theory on cryptographically secured chains as well as ideas for practical implementation.

2004

- Hal Finney introduces reusable PoW, a mechanism for receiving non-fungible tokens.

2008

- Satoshi Nakamoto publishes the Bitcoin white paper.

2009

- The Bitcoin Genesis Block is mined.
- The first bitcoin transaction occurs in block 170 between Satoshi Nakamoto and Hal Finney, the latter of whom receives 10 bitcoins from Nakamoto.
- The first Bitcoin exchange, called the New Liberty Standard, is established. Another exchange, Bitcoin Market, launches the same month.
- Nakamoto launches the Bitcointalk forum.

2010

- Laszlo Hanyecz spends 10,000 bitcoins on two Papa John's pizzas. Still, in later interviews, he's said he doesn't regret his $500m decision.

2011

- Bitcoin reaches $1 USD per coin for the first time.
- Charlie Lee launches Litecoin, which is a fork of Bitcoin.
- bidDNS launches on the Namecoin blockchain and allows users to mint unique, transferable domain names, hence serving as an initial application of NFTs.

2012

- The first bitcoin halving happens.

- Peercoin publishes a paper that introduces Proof of Stake.
- Coinbase raises $600k and launches its exchange platform.
- The Bitcoin Foundation is established to promote Bitcoin.
- Mihai Alisie and Vitalik Buterin (founder of Ethereum) launch *Bitcoin Magazine.*
- Chris Larsen and Jed McCaleb found OpenCoin, which leads to the development of the Ripple (XRP) transaction protocol.
- Meni Rosenfeld publishes a paper titled "Overview of Colored Coins" which details a small denomination of bitcoins that can be used to represent individual assets, hence serving as an early example of non-fungible token applications.

2013

- Bitcoin hits a $1bn market cap.
- Vitalik Buterin introduces Ethereum and smart contracts in a white paper.
- 25,000 bitcoins are stolen from a Bitcoin Forum founder's wallet.
- Dogecoin launches.
- GameKyuubi accidently types "I AM HODLING" in a Bitcointalk forum, hence turning "hodl" into a viral and quasi-official term.
- Coinbase raises $25 million in a Series B funding round.
- The FBI shut down the Silk Road and confiscates 26,000 bitcoins.
- Ripple (XRP) launches.

2014

- Kevin McCoy mints the first one-off NFT called "Quantum" and describes it as a monetized graphic.
- Microsoft accepts bitcoin as payment for Xbox games and Window software.
- Counterparty launches, which allows non-fungible asset creation.
- Dash (DASH) launches.
- Stellar (XLM) launch.
- Tether (USDT) launches as a stablecoin, which ties itself to the value of the U.S dollar.
- Hackers steal $450 million in bitcoins from the Mt. Gox exchange.

2015

- NASDAQ launches a blockchain trial.
- Coinbase raises $75 million in a Series C funding round.
- The Ethereum Frontier network launches, which enables developers to write smart contracts and dApps that can be deployed to a live network.
- NASDAQ initiates a blockchain trial.
- The Linux Foundation launches the Hyperledger project, which offers open source blockchain tools.
- 40 financial institutions in the US form the R3 consortium to explore how blockchain could benefit their operations.
- Ethereum Class launches.
- Etheria, the first NFT project, launches at Ethereum's first developer conference.
- The game creators of Spells of Genesis pioneer the issuance of in-game non-fungible assets and do so via Counterparty.

2016

- The DAO (Decentralized Autonomous Organization) loses $50m in a hack.
- The second Bitcoin halving happens.
- Bidorbuy, the largest South African marketplace, launches bitcoin payments for buyers and sellers.
- Bitfinex, a major crypto exchange, is hacked and 120,000 bitcoins are stolen.
- A bug in the Ethereum DAO code is exploited, leading to a hard fork of the Ethereum network.
- Counterparty teams up with a trading card game called Force of Will to launch tokenized cards.
- Metamask launches.
- Brave, an open-source privacy-focused web browser with a companion token (BAT) launches.

2017

- Bitcoin exchange Bitthumb is hacked to the tune of $20m.
- Virtual currencies are officially recognized in Japan and bitcoin becomes a legal payment method.

- Russia announces plans to legalize the use of cryptocurrencies, and such plans were later enacted in 2020.
- Bitcoin hits $10,000 USD per coin.
- Bitcoin Cash, another Bitcoin hard fork, launches with an increased blocksize limit.
- European banks form the Digital Trade Chain consortium to develop a trade finance platform based on blockchain.
- The Block.one company introduce the EOS blockchain operating system and token, aiming to support commercial dApps.
- Cardano (ADA) launches.
- The ERC-721 standard is proposed.
- The CryptoPunks project launches.
- The CryptoKitties project launches and goes viral, raising $12.5 million.
- Decentraland (MANA) launches.
- The OpenSea NFT marketplace launches.

2018

- Bitcoin turns 10.
- Switzerland accepts some tax payments in Ether and Bitcoin.
- South Korea requires that all Bitcoin traders reveal their identity.
- Google, Facebook, and Twitter ban the advertising of crypto projects.
- The European Commission launches the Blockchain Observatory and Forum.
- Decentraland raises $26 million in an ICO.
- The NFT-based game Axie Infinity launches.

2019

- Microsoft launches a blockchain service.
- Amazon opens up the Amazon Managed Blockchain service on AWS.
- Walmart launches a supply chain system using Hyperledger.
- Nike patents a system called CryptoKicks that uses NFTs to verify the authenticity of physical shoes.

2020

- The Frankfurt Stock Exchange admits the first bitcoin exchange-traded note.

- Paypal allows users to buy and sell a select number of cryptocurrencies on its platform.
- Ethereum launches the Beacon Chain in preparation for Ethereum 2.0, which aims to transfer to PoS.
- Dapper Labs releases the beta version of NBA TopShot, which sells NBA-related digital collectibles.

2021

- Shiba Inu (SHIB) experiences a 98.1 million percent increase and passes Dogecoin in market cap, fueled by a meme coin resurgence.
- CryptoPunks passes a billion-dollar valuation.
- Digital artist Beeple sells an NFT for a record-breaking $69.3 million.
- Coinbase goes public and ends its first day at an $85.7 billion valuation.
- Bitcoin surpasses a $1 trillion market cap.
- The first Bitcoin ETF (ticker: BITO) launches on the American stock market.
- El Salvador adopts Bitcoin as legal tender, becoming the first country to do so.

Legality

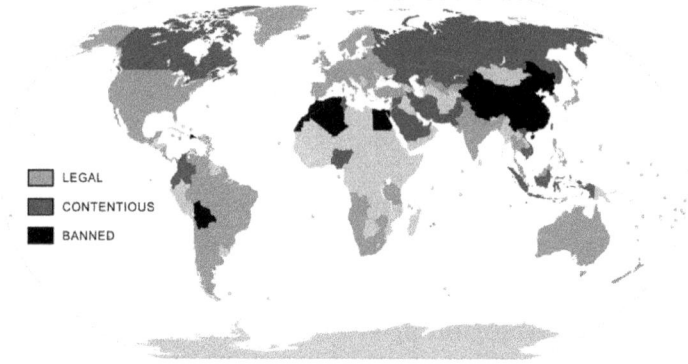

Cryptocurrencies are fully legal in all countries not on the following list as of early 2022.

Illegal:

Algeria

Egypt

Morocco

Bolivia

Nepal

China

Banking ban:

Nigeria

Canada

Columbia

Ecuador

Russia

Saudi Arabia

Jordan

Qatar
Iran
Bangladesh
Taiwan
Cambodia

Other:

Turkey - banking ban and illegal as payment tool.

Indonesia – illegal as payment tool.

Vietnam – illegal as payment tool.

Crypto Q&A

This section answers 15 of the most-asked questions about Bitcoin, Ethereum, and Blockchain. The questions progress as follows:

Bitcoin

- What is the point of Bitcoin?
- Is Bitcoin mining profitable?
- Is Bitcoin outdated?
- Is Bitcoin cyclical?
- Can you buy things with Bitcoin?

Ethereum

- Is Ethereum a scam?
- What is Ethereum versus Ether?
- Is Ethereum anonymous?
- Can Ethereum be hacked?
- Can the rules of Ethereum change?

Blockchain

- What is a Node?
- How do Smart Contracts work?
- What are Forks?
- Who invented Blockchains?
- What are Blockchains used for?

What is the point of Bitcoin?

The primary value of Bitcoin comes from its applications as a store of value and a means of private, global, and secure transacting. This, in essence, is the point of Bitcoin; a purpose which has been executed upon quite successfully, given its historical returns and the 300,000 or so daily transactions executed upon its network.

On a more ideological basis (Nakamoto would certainly agree with this), Bitcoin has served as a massive catalyst for all things decentralization, blockchain, and cryptocurrency. The impact of Bitcoin on the world, as well as its purpose, cannot just be measured in market cap or transaction volume, but is better viewed as an invention.

Inventions radically alter the world and spawn new ideas, communities, and ways of going about life. The idea of Bitcoin is the idea of a system that operates in a transparent and distributed manner. This, at its core, and not even examining its pros and cons, is a radically new idea, backed up by working technology, that offers an unprecedented alternative to communication and collaboration.

So, the technology of Bitcoin has a clear purpose, which is to serve as a distributed and trustless digital currency. The idea of Bitcoin also has a clear purpose, for all its downfalls—an idea that provides an alternative to most aspects of modern life and serves to incur and inspire a generation of builders, thinkers, and innovators to take their skills down a decentralized path in the name of a better world.

Is Bitcoin mining profitable?

It certainly can be. The average annual return on investment for those who rent the computational power (the rigs and nodes) of other miners can vary from high single-digits to low double-digits, while the annual ROI for self-managed Bitcoin mining varies throughout the double-digits. To put a number on gross returns, 40% to 200% annually can be expected, while 50% to 150% is normal. Net profits are largely determined by hardware cost and the price of the mined asset. Either way, this return beats the average historical stock market and real estate returns of approximately 10%. Still, Bitcoin mining is volatile and expensive, and a swath of factors influence each individual's returns. We'll take a moment to identify several of these factors.

Cryptocurrency Price. The major influencing factor in mining profit is the price of the given cryptocurrency asset. A 2x rise in bitcoin price results in 2x the mining profit (because the amount of bitcoin being earned stays the same, while the equivalent value changes, assuming no difference in difficulty), while a 50% drop results in half the profits.

Most miners must sell earnings to cover costs, which renders price action (especially given the volatile nature of cryptocurrencies and especially that of bitcoin) extremely important. If a miner can afford not to sell and holds for the long-term, short-term price action becomes less important.

Hash Rate and Difficulty. Hash rate is the speed at which equations are solved and blocks are found. Hash rates for miners roughly equate to earnings, and more miners entering the system (thus increasing the hash rate of the network and the related mining difficulty, which is a term that describes how hard it is to mine blocks) dilutes per-miner hash share and therefore profitability. In this way, competition drives down profit through difficulty and hash rate.

Price of Electricity. As the mining process becomes more difficult, electricity requirements also increase. The price of electricity plays a major part in profitability and, apart from politics, largely determines the geographic placement of industrial mining farms.

Halving. Every 4 years, the block rewards programmed into Bitcoin halve to incrementally reduce the influx of coins. Since May of 2020, miner rewards have been 6.25 bitcoins per block. In 2024, block rewards will fall to 3.125 bitcoins per block, and so on every four years thereafter. In this manner, long-term mining rewards must fall unless the per-coin value rises as much or more than the equivalent decrease in block rewards, or difficulty decreases in equivalence to reward decreases.

Hardware Cost. Of course, the actual price of the hardware needed to mine bitcoin plays an extremely large part in profit and ROI. Mining can be set up easily on normal PC's (if you have one, check out nicehash.com); that said, setting up full rigs involves the cost of motherboards, CPUs, graphics cards, GPUs, RAM, and more. The simplest way to set in a mining station is to buy pre-made rigs, but this involves paying a hardware premium. Building your own rig saves money, but also requires some technical knowledge. Generally, do-it-yourself options cost at least a few thousand dollars (including graphic cards), while pre-made rigs cost, at minimum, a bit more. Since graphic cards require the bulk of the investment, make sure to scour local resale websites for deals, as retail prices for such cards have gone through the roof since the advent and growth of mining.

To conclude this question, the variables influencing mining profitability are numerous and subject to rapid change, and potential earnings are biased towards large farms with access to cheap electricity. That said, crypto mining is still very much profitable and returns likely will, for some time, remain far ahead of normal returns in most other asset classes. This excludes the potential of an extended market-wide collapse or the advent of harsh governmental regulation.

Is Bitcoin outdated?

Yes. The technology powering Bitcoin is outdated relative to newer competitors. Bitcoin did the work of trailblazing and acted as a proof-of-concept for cryptocurrencies, but as with all technology, innovation pushes forwards and keeping up with such innovation requires cohesive and major upgrades, which Bitcoin hasn't had.

The Bitcoin network can handle about 7 transactions per second, while Ethereum (the second-largest cryptocurrency by market cap) can handle 30 transactions per second and Cardano, the third largest and much newer cryptocurrency, can handle about 1 million transactions per second. Network congestion on the Bitcoin network leads to higher fees, and in this way, as well as in programmability, privacy, and energy use, Bitcoin is somewhat outdated.

This is not to say Bitcoin doesn't work; it does, it just means either serious upgrades should be implemented, or user experience will worsen, and competitors will thrive over time. Regardless, Bitcoin has enormous brand value, a massive scale of usage and adoption, and protocols that get the job done in a secure manner.

We'll likely see neither the best (large upgrades that solve all problems) nor worst (too outdated, no longer works, and goes to zero) situation play out; instead, the middle ground is likely, in which Bitcoin continues to face problems, continues to implement solutions, and continues to grow (although growth will certainly slow over time) in parallel to the crypto space.

Is Bitcoin cyclical?

Yes. Bitcoin is historically cyclical and tends to operate on multi-year cycles that generally consist of a breakthrough high, a correction, accumulation, recovery, and continuation. This can be simplified into a big up, major down, little up or sideways, and a big up.

Breakthrough highs have historically happened within a year of Bitcoin's halving events, which happen every four years. This, by no means, is an exact science, but it does provide some perspective on the medium-term potential and price action of bitcoin.

Additionally, large jumps of altcoins (specifically medium-to-small altcoins) typically occur while bitcoin is neither making a major upwards move nor a major downwards move, and often follow a large upward move. This occurs as investors take profits out of bitcoin (while the price consolidates) and puts them into smaller coins.

So, all this is generally something to think about, especially if you're thinking about trading bitcoin as opposed to holding for the long-term.

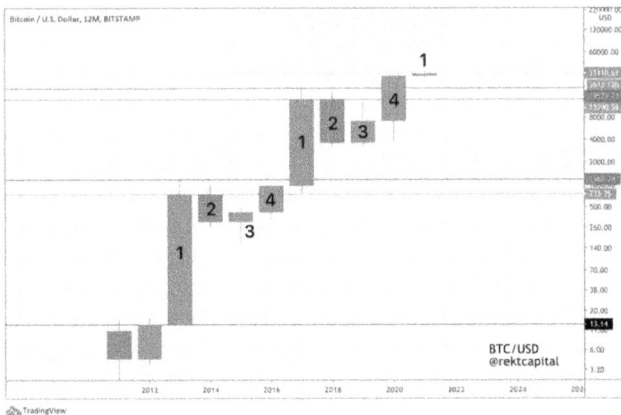

---

[14] Image credit to hackernoon.com. Take note that cycles are thought to be lengthening, which will serve to play a part in decreasing overall volatility as Bitcoin matures.

Can you buy things with Bitcoin?

Bitcoin represents a shared sense of value; value can be transacted and exchanged for equivalent value, just like any other currency. Despite this, it's quite difficult to directly buy most things with bitcoin, although options do exist and are rapidly multiplying.

Of course, one may always just exchange bitcoin for their given currency and use such currency to buy things, but the question remains: why can you not yet use bitcoin to purchase any items you would otherwise pay for with other digital payment methods? Such a question is complex, but mostly has to do with the fact that the established system of government-backed currencies has worked for quite a while, while cryptocurrencies are new, have yet to integrate into commercial systems, and operate outside of direct government control.

Current trends point to cryptocurrencies integrating to a much greater extent into online (and offline) retailers, wholesalers, and independent sellers. Already, Microsoft (in the Xbox store), Home Depot (via Flexa), Starbucks (via Bakkt), Whole Foods (via Spedn), and many other companies accept bitcoin; the tipping points will occur at the time of major online retailers accepting bitcoin (Amazon, Walmart, Target, etc.) and the time at which governments either embrace or push back against cryptocurrencies as a payment method.

The next page lists companies, which have at some time or in some location, allowed bitcoin as a payment method. Data is sourced from icoholder.com and is extremely liable to change.

- Alzashop.com
- Archive.org
- Badoo.com
- BigFishGames.com
- Menlo Park
- Bitcoincoffee.com
- Bloomberg.com
- Braintree
- CheapAir.com
- Crowdtilt.com
- Dish Network
- Domino's Pizza
- eGifter.com
- Etsy.com
- EZTV
- Euro Pacific
- Expedia.com
- ExpressVPN.com
- Fancy.com
- Grooveshark
- Gyft
- Intuit
- Lumfiles
- Mega.co.nz
- Microsoft
- Mint.com
- Namecheap.com
- REEDS Jewelers
- OKCupid
- Overstock.com
- Paypal
- PizzaForCoins.com
- Rakutan
- Red Cross
- Reddit
- Seoclerks.com
- Shopify.com
- Stripe
- Subway
- The Pirate Bay
- Virgin Galactic
- Watchshopping.com
- Wikipedia
- WordPress.com
- Zynga.com
- Newegg.com
- Burger King
- Lieferando
- Subways
- Cheapair.com
- Royal Bank of Canada
- Bankera
- University of Columbia
- Intuit

Is Ethereum a scam?

While Ethereum itself is not a scam and receives some degree of governance from the Ethereum Foundation and the robust group of individuals that participate in the platform, it is possible for others to use Ethereum to run scams on unsuspecting investors. Here are a few scams that can happen within cryptocurrency markets:

Users may find "airdropped" tokens in their wallet that were placed there by the scammers. When a user goes to a decentralized exchange (DEX) to trade the tokens, they have to approve access to their wallet for that token. Without reviewing the "contract", they sign off on it via their wallet. The permission they signed off on also allowed access to other coins or tokens in the wallet, which get traded away and stolen, leaving the user with worthless tokens.

Initial Coin Offering (ICO) scams were more common in the earlier days of the token economy. This would involve a hyped-up company doing interviews, having a top-quality website, and selling a token on the network. After the ICO is over, the token becomes worthless, but the company that held the ICO has all of the Ether, bitcoin, or other coins that were paid to them for the tokens. The SEC has since started pursuing and prosecuting such individuals and organizations.

A "rug pull" involves the release of a token, typically with some hype, on DEXs. The originators of the token sell their holdings and remove the liquidity required to allow trades to occur. This leaves the rest of the token participants holding the bag with no effective way to trade out of their position. This has become a more common scam and requires careful consideration and research before taking a position in a token.

DeFi farming is a fairly new trend that involves staking tokens on a farm. The farm will reward stakeholders with some other token (usually branded with the farm name, such as KRILL or MOON) based on their stake size. Occasionally these farms are scams, and the farm creators can make off with some or all of the staked tokens.

Alternately, the farm may be hacked due to poor security practices on the part of the creators, and some third party may make off with the tokens.

As a last note, it pays to research and verify the validity of where you put your coins or tokens. Because there is so little regulation and so much transparency in the community, you are liable for your own mistakes. Sticking to mainstream exchanges and coins is the safest strategy, and investments into more pioneering spaces of the crypto world should be taken only after thorough research.

What is Ethereum versus Ether?

Ethereum is the term used to describe the network, platform, and overall ecosystem of the network. It is also the name of the foundation that provides governorship and technical guidance to the network.

Ether is the actual coin produced on the network, and the coin used for transaction fees incurred by any action that takes places on the network—including the sending of any token supported by the network.

Is Ethereum anonymous?

The core nature of Ethereum and other cryptocurrencies is inherently anonymous. In theory, a person can mine Ethereum, hold it, swap to various tokens within the Ethereum ecosystem, and swap back to Ethereum completely anonymously.

The anonymity becomes more complex when working with services external to Ethereum, such as exchanges. Because of government regulation, almost all exchange services require registration and some form of identity verification in order exchange cryptocurrencies for fiat (government) currencies. This aids the exchanges in compliance with tax law, as cashing out cryptocurrency typically requires some form of tax to be paid to the government.

As such, Ethereum itself is anonymous, but interfacing with external resources involving Ethereum is typically not anonymous.

Can Ethereum be hacked?

Because Ethereum operates on an open and exposed blockchain, it is not so easy to hack. Possible attacks include 51% attacks, brute-force attacks against private keys, and hack attempts against centralized exchanges.

Centralized Exchanges

Centralized exchanges have their own software, often developed in-house, that handles the high-speed interactions needed to trade cryptocurrency effectively. These exchanges require you to deposit funds for trading, and those funds are very much within the control of the exchange, and in the exchange's own wallet.

Because exchanges do not typically use any open-source software, there is not a robust auditing process around their networks or configurations. If there is a suitable cybersecurity weakness in the exchange platform, it is absolutely possible for theft to occur.

Internal exchange employees, too, may have credentials that permit them to engage in theft. Exchange users with weak credentials, or credentials that have been leaked from other hacks, may also become victims of theft, though such theft would be limited to a smaller group of individuals.

Brute-Force Attacks

Because most cryptocurrency wallets exist as a private-key/public-key pair, and because the process for generating the public key and wallet address from the private key is a known method, it is possible for clever hackers to brute-force a wallet with a balance and transfer those funds somewhere else.

The main defense against this process is the sheer scope of private keys available to use, and the fact that every wallet's private key is chosen at random (except in very specific cases). A private key is two hundred and fifty-six bits long. Represented in hexadecimal, private keys appear in this format:

8da4ef21b864d2cc526dbdb2a120bd2874c36c9d0a1fb7f8c63d7f7a8b41de8f

Each digit of the private key can represent sixteen different values (0-9, A-F). A brute force attack works off a single address or list of public addresses and can be performed on a CPU or GPU for higher speed. The brute force software generates private key values and runs them through the address generation process to see if they match. If a match is found, the private key is output to the screen or dumped to a file. If no match is found, the system continues searching.

Modern processors can perform this function at over one million keys per second per core, while graphics processors can perform these functions at over two billion keys per second assuming the system is tuned properly. While this sounds like a lot and may make one concerned over the security at the core of cryptocurrency, it is in practice not at all a threat. Because of the sheer number of possibilities (the number of private keys available is greater than atoms in the universe) it would take an absolutely impossible amount of time, statistically speaking, to find even one wallet—longer than the longest estimates of the age of the earth.

While this bodes very well for the security of cryptocurrency, it emphasizes the importance of retaining your wallet backups or mnemonic phrases. If lost, nobody can recover your wallet for you.

Can the rules of Ethereum change?

The Ethereum network has many rules and protocols that define its function; from tokens and smart contracts to block rewards and many other specific functions that govern the operation of the network. The developers of Ethereum built the software to be flexible and implementation of new rules is possible. The Ethereum foundation has processes to propose and implement improvements or changes to the platform using Ethereum Improvement Proposals (EIPs). This process allows for thoughtful and careful updates to the functions of the network. Some of the more famous EIPs involve new network functions:

EIP-20 proposed the inclusion of the ERC-20 token standard. This EIP alone has enabled the entire token ecosystem, initial coin offering (ICO) model, and decentralized exchanges. The ability to create a cryptocurrency without creating an entire blockchain has massively altered the landscape and lowered the barriers to adoption.

EIP-720 proposed the creation of the ERC-721 non-fungible token (NFT) standard. This standard has enabled an entirely new model of representing objects with value on the blockchain. NFTs were the industry buzzword in 2021, with decentralized exchanges popping up and allowing users to purchase and own NFTs of artwork, music, and other things. Further development of these concepts could lead to real-life integration of physical objects as NFTs.

What is a Node?

A node is a computer connected to a blockchain's network which assists the blockchain in writing and validating blocks. Some nodes download an entire history of their blockchain and perform more tasks than regular nodes. Additionally, nodes are not tied to a specific network; nodes can switch to different blockchains practically at will, as is the case with multipool mining. In the Ethereum ecosystem, there are three different types of nodes that can be run, and most other cryptocurrencies follow a similar model.

Light Node: A light node downloads and stores only the header chain and makes requests of the network when other chain components are needed. A light node is capable of verifying data in the block headers.

Archive Node: Archive nodes store the full chain and use it to build an archive of historical states. Normally, if you want to view the state of the chain at a given block, a separate application will need to process and analyze all blocks up to the block in question. Running an archive node essentially allows for a time-machine view of all accounts at all times in the chain history. This is storage intensive, and suitable mostly for those looking to run their own block explorer or other analytics services.

Full Node: This type of node stores the full blockchain, participates in block validation and verification, and provides network services to other Ethereum clients. While a full node does not store historical states of the blockchain, they can be computed from a full node through analytic tools.

Running an Ethereum node local to your own Ethereum-interfaces applications can be highly beneficial in terms of ensuring the lowest possible latency for transactions. Because the node is being run locally, you can trust that it is secure, and you don't necessarily need to trust a third-party node. You can also choose to ensure redundancy against things like hardware failures or power outages, knowing in full how your system is configured. This is not possible with a third-party node that may go offline unexpectedly.

How do Smart Contracts work?

A smart contract is a scripted program that runs on the Ethereum network. Rather than an endpoint on the network being responsible for storing and running the program, the entire network supports the function of the contract as it interacts with the blockchain.

They exist as a type of Ethereum account, which means that a smart contract has a wallet address, balance of coins or tokens, and can send and receive transactions programmatically without any user engagement.

As users submit coins or tokens to the contract address, pre-programmed functions can take effect and perform any number of activities. Once a smart contract has been defined on the network, they cannot be deleted, and interactions with them are not reversible, e.g., you cannot request a refund of ETH sent to a contract address.

A simple smart contract example would be if you sent an amount of ETH to the contract address. The contract would register the receipt of the ETH and automatically send you an equally valued amount of WETH (wrapped ETH, a token on the Ethereum blockchain that hold equal value with ETH).

A smart contract can do substantially more than a basic coin/token swap. Deflationary mechanisms, airdrops, redistribution of transaction fees, and much more are able to be instilled into the blockchain and function autonomously through smart contracts.

What are Forks?

A fork is the occurrence of a new blockchain being created from another blockchain. Bitcoin has had 105 forks, the largest of which is the present-day Bitcoin Cash. Forks occur when an algorithm is split into two different versions.

Two primary types of forks exist: hard forks, which occur when all the nodes in the network upgrade to a newer version of the blockchain and leave the old version behind (two paths are then created: the new version and the old version), and a soft fork, which contrasts this by rendering the old network invalid (resulting in just one blockchain). Flip back to page 134 for a visual of the forking process.

Who invented Blockchains?

Blockchain has a history spanning back to the 1990s, at which time a cryptographically secured chain of blocks was conceptualized for the first time by Stuart Haber and W. Scott Stornetta. More papers on cryptographically secured chains were published in 2000, and in 2009 Satoshi Nakamoto implemented the first blockchain in the Bitcoin network.

Bitcoin wasn't just the first successful application of blockchain, but also solved several problems that had previously impeded implementation. The success of Bitcoin was followed up by altcoins, which improved upon the original concept for blockchain by adding features such as smart contracts, proof of stake, and various scaling solutions, such as layer 2 protocols, sharding, ring signatures, and the lightning network.

Implementation of these technologies and more, combined with an ever-growing number of use cases, are incurring the next generation of blockchain-backed technologies, backed by coins such as Cardano, Solana, Binance, Ripple, and more.

What are Blockchains used for?

Bitcoin was the first successful large-scale use case for blockchain, but blockchain as a technology goes far beyond Bitcoin. Here are fields in which blockchain will have or is having a major effect:

- Digital currencies
- Logistics, data, and supply chain management
- Cross-border payments and means of transaction
- Artist royalty tracking
- Secure storing and sharing of medical data
- NFT marketplaces
- Voting mechanisms
- Verifiable ownership of real estate
- Invoice reconciliation and dispute resolution
- Ticketing
- Financial guarantees
- Proof of insurance/insurance policies
- Health/personal data records
- Capital access
- Decentralized finance
- Digital identity
- Process/logistical efficiency
- Data verification
- IP protection
- Digitization of assets and financial instruments
- Governmental financial transparency
- Online gaming
- Syndicated loans
- And more!

Popular Cryptocurrencies

Bitcoin, Ethereum, and Litecoin are excluded from this section given the ample coverage presented throughout the book. A look at the history and vision of top cryptocurrencies constitutes this section, presented in order of market cap:[15] Links are included to the website and whitepaper of each asset.

1. Binance Coin (BNB)
2. Tether (USDT)
3. Solana (SOL)
4. Cardano (ADA)
5. XRP (XRP)
6. Polkadot (DOT)
7. USD Coin (USDC)
8. Avalanche (AVAX)
9. Dogecoin (DOGE)
10. Terra (LUNA)
11. Uniswap (UNI)
12. ChainLink (LINK)
13. Polygon (MATIC)
14. Algorand (ALGO)
15. Elrond (EGLD)
16. Axie Infinity (AXS)
17. Decentraland (MANA)
18. Tron (TRX)
19. Cosmos (ATOM)
20. Theta (THETA)

---

[15] As of early 2022. Market cap data is based upon circulating supply, not maximum supply.

Binance Coin (BNB)

The Binance exchange, named as such by the meshing of "Binary Finance", aimed to solve problems presented by earlier exchanges, those being poor technical architecture, limited security, poor customer service, and poor internationalization and language support.

The Binance exchange launched in 2017 with a matching engine capable of sustaining 1.4 million orders per second. At that time, only five trading pairs were supported. Binance has since exploded to $1 billion in profit in 2020 and 13.5 million users. The Binance coin (BNB) launched alongside the exchange in 2017 on the Ethereum network. The coin would be used to pay various exchange fees.

BNB eventually moved to its own blockchain called Binance Smart Chain (BSC) and grew to become the current third-largest cryptocurrency in existence with a market cap of just about $100 billion.[i]

Visit Binance at binance.com or binance.us and their whitepaper at exodus.com/assets/docs/binance-coin-whitepaper.pdf

Tether (USDT)

Originally launched as Realcoin in 2014, and renamed in November of that year, Tether aimed to provide a digital token backed by fiat currency to enable a decentralized method of exchanging value while using a familiar accounting unit. This unit had to be less volatile than most other cryptocurrencies, and Tether achieved price stability by maintaining a one-to-one reserve ratio between all Tether tokens and the associated fiat currency. Tether became the first successful stablecoin and remains pegged to the US dollar.

Tether's stability has since provided a more viable means of exchange. Say you're trying to buy lunch—if you pay in bitcoin, the merchant doesn't know what that bitcoin will be worth in a day, a week, or a month. The inherent volatility (fees aside) makes transacting difficult. Stablecoins solve this problem by associating with the

US dollar or any other stable reserve asset, hence creating an all-around much more pleasant and familiar transaction environment.

Tether is now an ERC20 token (meaning it runs on the Ethereum blockchain) and the fourth-largest cryptocurrency. It currently presents an astounding 24hr volume of $85 billion.[ii]

Visit Tether at tether.to and their whitepaper at tether.to/wp-content/uploads/2016/06/TetherWhitePaper.pdf

Solana (SOL)
Solana, launched in March of 2020, aimed to provide an alternative to Ethereum by utilizing a blockchain architecture based upon Proof of History (PoH), which can verify order and passage of time between events (data) in blockchains. PoH is used alongside the consensus algorithm Proof of Stake (PoS).

While Ethereum can support between 15 and 45 transactions per second, Solana can handle tens of thousands, and Solana fees are also significantly lower than those of Ethereum. Since launch, Solana has skyrocketed to a $67 billion market cap and a reputation as being the coin, if any, that could challenge Ethereum hegemony.[iii]

Visit Solana at solana.com and their whitepaper at solana.com/solana-whitepaper.pdf

Cardano (ADA)
Much like Solana, Cardano was designed to improve upon the many features of Ethereum. Launched in 2015 by Ethereum co-founder Charles Hoskinson, Cardano aimed to provide a more balanced and sustainable ecosystem and a standards and peer-driven approach to research and improvement.

Cardano runs on its own blockchain and utilizes a proof of stake protocol called Ouroboros, which runs protocols similar to those of Google and Facebook and

174

presents scaling solutions far beyond those of Bitcoin and other cryptocurrencies. Cardano has grown into a $59 billion market cap and aims to "connect the unconnected" through ambitious plans to reach a billion people.[iv]

Visit Cardano at cardano.org and their whitepaper at whitepaper.io/document/581/cardano-whitepaper.

Ripple (XRP)

Ripple, launched in 2012 by Ripple Labs, built upon a novel consensus algorithm (RPCA) that presented short wait times and low fees relative to coins such as bitcoin.

Ripple reported interest from 100+ banks in using its XCurrent messaging technology, and some even planned to use the XRP coin. However, in 2018, the U.S Securities and Exchange Commission (SEC) filed a class action against Ripple stating that the company had raised hundreds of millions through unregistered sales of XRP tokens, which led to exchange delistings and public pushback.

Still, Ripple has largely recovered, and sits as the seventh-largest cryptocurrency with a market cap of $50 billion.[v]

Visit XRP at ripple.com and their white paper at ripple.com/files/ripple_consensus_whitepaper.pdf

Polkadot (DOT)

Launched in 2020 by the Web3 Foundation, Polkadot aimed to improve upon past coins in five areas: scalability, isolatability, developability, governance, and applicability. The Polkadot network protocol is designed to be a general-purpose blockchain and utilizes pooled security and trust-free interchain transactability to enhance scalability.

The DOT token originally operated through Proof of Authority (PoA) and switched to Nominated Proof of Stake (NPoS), which selects validators to participate in consensus protocol (the staking). Polkadot is currently the eighth largest cryptocurrency and maintains a market cap of $40 billion.[vi]

Visit Polkadot at polkadot.network and their whitepaper at polkadot.network/PolkaDotPaper.pdf

USD Coin (USDC)

USD Coin was launched in 2018 by Circle and the Centre consortium on the concept of improvising upon Tether (the first large-scale stablecoin) by creating a fiat-collateralized stablecoin with strong governance and transparency.

Each dollar is backed by "fully reserved assets" and USDC can tokenize US dollars through a process involving US dollars being sent to a coin issuer's bank account, the issuer providing a smart contract to create an equivalent amount of USDC, and the minted coins being sent to the user with the dollars held in reserve.

Visa announced that transactions could be settled using USDC in March of 2021 and USDC currently sits at a $38 billion market cap.[vii]

Visit USD Coin at centre.io/usdc and their whitepaper at f.hubspotusercontent30.net/hubfs/9304636/PDF/centre-whitepaper.pdf

Avalanche (AVAX)

Introduced in 2018, Avalanche launched with the aim of building a "high-performance, scalable, customizable, and secure blockchain platform" which quickly became a leading Ethereum competitor. Users can build and launch decentralized applications (dApps) on the Avalanche platform.

As the fastest smart contract platform in the blockchain industry, the Avalanche network has grown to hundreds of dApps and a $26 billion market cap.[viii]

Visit Avalanche at avalabs.org and avalabs.org/whitepaper

Dogecoin (DOGE)

Dogecoin was conceptualized by Billy Markus and Jackson Palmer, a pair of engineers from (respectively) IBM and Adobe. Palmer conceptualized Dogecoin, while Markus designed the DOGE protocol based on Litecoin.

The coin was originally meant to be nothing more than a joke, yet it launched in December 2013 and jumped by hundreds of percentage points in its first month. In 2014, Dogecoin trading volume briefly suppressed that of every single cryptocurrency (including Bitcoin) combined. Throughout 2020 and 2021, social hype pushed Dogecoin up thousands of percent, second only in meme coin rankings to Shiba Inu.

The Dogecoin community continues to build utility around the coin and serves as an interesting example of psychology and generational splits in financial perspectives. DOGE peaked at a $70+ billion market cap and currently sits at $28 billion.[ix]

Visit Dogecoin at dogecoin.com.

Terra (LUNA)

Launched in 2019, Terra aimed to achieve price-stability via an elastic money supply and stable mining incentives, and in this way create a cryptocurrency that combined the best of the decentralized nature of Bitcoin and the relative stability of fiat currencies. Terra launched behind the Terra Alliance, which was a collective of 15 e-commerce companies in Asia that annually process 25 billion USD in transaction volume. The Alliance would provide the necessary scale to fuel Terra adoption.

Today, the Terra protocol serves as a stablecoin network within which users can create stablecoins pegged to fiat currencies, while the LUNA coin serves as the reserve asset. LUNA sits at a $16 billion market cap.

Visit Terra at terra.money and their whitepaper at assets.website-files.com/611153e7af981472d8da199c/618b02d13e938ae1f8ad1e45_Terra_White_paper.pdf

Uniswap (UNI)

Uniswap (originally named Unipeg, as a mixture between a unicorn and pegasus) launched in 2018 with the intent of fully embracing the properties of being censorship resistant, decentralized, permissionless, and fully secure, which the inventor (Hayden Adams) felt to be lacking in other major projects on the Ethereum network. The Uniswap exchange launched as a simplistic smart contract interface for trading ERC20 tokens, and the UNI token launched in 2020.

The Uniswap exchange passed $500 billion in cumulative trading volume in 2021 and UNI sits at a 12 billion market cap.[x]

Visit Uniswap at uniswap.org and their whitepaper at uniswap.org/whitepaper.pdf

ChainLink (LINK)

Created in 2014 and launched in 2017, ChainLink developed Decentralized Oracle Networks (DONs), which are networks maintained by a committee of [ChainLink] nodes that pair with smart contracts, ultimately providing extremely secure transactions.

DONs allow ChainLink to develop hybrid smart contracts, eliminate complexity, enable scale and confidentiality, and pursue security through strong incentive mechanisms. The ChainLink network has since secured $83 billion worth of value and LINK sits a $10 billion market cap.[xi]

Visit ChainLink at chain.link and their whitepaper at chain.link/whitepaper.

Polygon (MATIC)

Launched in 2017 on the Ethereum blockchain, Polygon aimed to solve several problems faced by Ethereum, those being low throughput, high fees, and lack of sovereignty. Polygon launched as "a protocol and framework for building and connecting Ethereum-compatible blockchain networks" and switched to the Polygon blockchain (as opposed to the Ethereum blockchain) in 2019.

Now, Polygon works as a layer 2 scaling solution for Ethereum-compatible networks—even surpassing Ethereum in user base in October 2021.[16] MATIC sits at an $18 billion market cap.[xii]

Visit Polygon at polygon.technology and their whitepaper at polygon.technology/lightpaper-polygon.pdf

Algorand (ALGO)

Launched in 2019, Algorand aimed to provide improved security, scalability, and fast transaction finality in a completely permissionless environment through a novel Byzantine Agreement Protocol.

Today, the Algorand network has 15 million addresses and achieves block finality in 4.39 seconds. ALGO sits at an $11 billion market cap.[xiii]

Visit Algorand at algorand.com and their whitepaper at algorandcom.cdn.prismic.io/algorandcom%2Fece77f38-75b3-44de-bc7f-805f0e53a8d9_theoretical.pdf

Elrond (EGLD)

Founded in 2017, Elrond aimed to create a highly scalable public blockchain through a combination of Adaptive State Sharding and Secure Proof of Stake (SPoS). Elrond's architecture solves the energy and computational waste problems of Bitcoin and Ethereum and ensures fair distribution with the novel SPoS model Elrond launched a wallet and payments app called Maiar in 2021 that aims to become the Venmo of crypto, and the EGLD coin sits at a $14 billion market cap.[xiv]

Visit Elrond at elrond.com and their whitepaper at elrond.com/assets/files/elrond-whitepaper.pdf

---

[16] According to Etherscan.

Stellar (XLM)

Stellar Lumens introduced a new consensus model called federated Byzantine agreement (FBA). The Steller Consensus Protocol (SCP), which built upon FBA, combines the decentralized control provided by PoW and PoS with the low latency and asymptotic security of Byzantine agreements.

Stellar received $3 million isn seed funding from Stripe in 2014 and grew to 3 million registered accounts in year one. 80+ projects now operate on the Stellar blockchain and XLM sits at an $8 billion market cap.[xv]

Visit Stellar at stellar.org and their whitepaper at stellar.org/papers/stellar-consensus-protocol?locale=en

Axie Infinity (AXS)

Axie Infinity launched in 2018 as one of the first crypto games, being a Pokémon-inspired universe in which players can earn tokens through gameplay and PVP battles, as well as by breeding "Axies" and trading them on the marketplace.

Axie Infinity reached 2 million daily active users in October of 2021 and sits at $3.4 billion in total sales volume, while the AXS token has an $8 billion market cap.[xvi]

Visit Axie Infinity at axieinfinity.com and their whitepaper at whitepaper.axieinfinity.com

Decentraland (MANA)

Decentraland is a blockchain-based virtual world (a metaverse) powered by the Ethereum blockchain. Land is permanently owned as NFTs by users, and the MANA token is used to purchase in-world goods and services.

Work began on Decentraland in 2015 and a beta version launched in 2017. At that time, land parcels sold for as little as $20 each, while a Decentraland property bundle sold just four years later for a record $2.43 million. MANA sits at a $10 billion market cap.[xvii]

Visit Decentraland at decentraland.org and their whitepaper at decentraland.org/whitepaper.pdf

Dai (DAI)

Formed in 2014 (the same year as the first stablecoin, Tether) the Maker Protocol and Dai token launched with the intent of becoming an "unbiased, collateral-backed cryptocurrency soft-pegged to the US Dollar" that minimized price volatility. The Dai Stablecoin System, later renamed to the Maker Protocol, remains equivalent to the US dollar through a series of dynamic feedback systems called collateralized debt positions (CDPs).

CDPs are smart contract derivatives on the Ethereum blockchain that represent a debt contract and are collateralized by crypto securities. Once money is deposited into the smart contract, the CDP allows the user to generate an equivalent USD value to the value in Dai they wish to borrow.

While this may seem complicated, Dai operates on the surface just like any other cryptocurrency (ignoring volatility) and offers the benefits of digital and decentralized currencies without the volatility and risk. DAI sits at a $6.5 billion market cap.[xviii]

Visit Dai at makerdao.com and their whitepaper at makerdao.com/en/whitepaper

Cosmos (ATOM)

Cosmos launched in 2014 with the mission of creating blockchains that could communicate and interact with each other, as opposed to the standard isolated networks. These interoperable blockchains can retain their security properties, and interoperability does not impede scalability.

The Cosmos network now consists of many independent blockchains, called zones, which are collectively powered by the Tendermint Byzantine fault tolerant (BFT) consensus engine. The ATOM coin sits at a $6 billion market cap.[xix]

Visit Cosmos at cosmos.network and their whitepaper at v1.cosmos.network/resources/whitepaper

The Sandbox (SAND)

The Sandbox aims to build a virtual world in which players can "build, own, and monetize their gaming experiences" using the SAND utility token. The Sandbox can be thought of as similar to Minecraft and Roblox, except players truly own their in-game property through NFTs.

In-game land is scarce, tradeable, and customizable, and players can "craft, play, share, collect, and trade without central control." The Sandbox also offers the Voxel Editor, which allows users to create and animate 3d objects, a marketplace, in which users can upload, publish, and sell their VoxEdit creations (to be used in-game), and a game maker, in which owned assets can be placed on owned land in the digital landscape.

The Sandbox alpha version went live in November 2021, and its 12,000 owners now collectively own $500 million worth of digital land.[xx]

Visit the Sandbox at sandbox.game and their whitepaper at https://installers.sandbox.game/The_Sandbox_Whitepaper_2020.pdf

Helium (HNT)

Helium is a decentralized network that "enables devices anywhere in the world to wirelessly connect to the internet." Helium aims to disrupt the monopolized wireless network industry and remove the need for a central intermediary in receiving wireless network coverage.

Decentralized wireless internet coverage works through a distributed network of miners, who provide wireless network coverage in return for HNT tokens. The Helium network has 300k+ live nodes, just recently partnered with Dish Network to provide distributed 5G access, and grew HNT to a $4.5 billion market cap.[xxi]

Visit Helium at helium.com and their whitepaper at whitepaper.helium.com

EOS (EOS)

EOS launched in 2018 as the offspring of the private blockchain company block.one (b1.com) with the intent of introducing a highly configurable and fast platform for blockchain innovation and performance, largely targeted at institutions and developers.

EOS famously raised $4.1 billion in its 2018 ICO, and currently sits at a $3.6 billion market cap.[xxii]

Visit EOS at eos.io and their whitepaper at github.com/EOSIO/Documentation/blob/master/TechnicalWhitePaper.md

Enjin Coin (ENJ)

Enjin is the largest gaming community creation platform, with 18.7 million registered users, and aims to provide a decentralized platform to manage, distribute, and trade virtual goods, as well as tools to create and enable new cross-application gaming and NFT experiences.

More than 250,000 gaming communities and thousands of games create the vibrant community of the Enjin network, and the Enjin Coin clocks in at a $3.3 billion market cap.[xxiii]

Visit Enjin at enjin.io and their whitepaper at cdn.enjin.io/downloads/whitepapers/enjin-coin/en.pdf

Gala (GALA)

The Gala Games ecosystem is a gaming platform that operates through 16,000 player-run nodes and provides voting mechanisms to determine which games should be funded and brought onto the Gala platform. Within the ecosystem, 90+ team members develop NFT and blockchain-based games. The GALA token sits at a $3 billion market cap.

Visit Gala at app.gala.games and their whitepaper at support.gala.games/en_US/gala/what-is-gala

Dictionary

Full Dictionary: A to Z

Following this point, the rest of the book is a dictionary containing every term mentioned prior in this book, as well as many not mentioned. Please keep the following in mind:

- Words are sorted alphabetically from A-Z. This excludes a select few that, for various reasons, have been inserted out of order. If you cannot find a word in the alphabetical spot that it should occupy, refer to the index at the end of the book.

- Some concepts, terms, or technical aspects have been simplified and/or removed in the name of comprehension. To understand terms and concepts at an advanced level, consult in-depth outside sources, many of which can be found within the resource section at the back of this book.

1m, 3m, 5m, 15m, 30m, 1h, etc

Periods of time, M meaning minutes, H meaning hours, and D meaning days. Commonly found on charts.

51% Attack

A 51% attack refers to an attack on a blockchain network, similar to a hack, though a more apt comparison is to a hostile takeover of a stock and the corresponding company. Attacks are carried out by a group of miners who control more than 50% of a network's computing power. Once the group controls more than 50% of the computing power, they can deliberately alter transactions, since blockchain consensus is majority based. Once they control the network, they may issue transactions twice (called double-spending) and pocket the profit. However, 51% attacks are largely theoretical and practically impossible to successfully carry out on most established cryptocurrencies.

Account

An account is a pair of public and private keys from which you can control your funds. Accounts are typically viewed through an exchange, which provides an ideal user interface. However, funds are actually stored on the blockchain, not in accounts.

Address

An address, also known as your public key, is a unique collection of numbers and letters that function as an identification code, comparable to a bank account number or an email address. With it, you can carry out transactions on the blockchain. Addresses have round, colorful "logos," that are called address identicons (or, simply, "icons"). These icons allow you to quickly see whether or not you input a correct address.

Airdrop

An airdrop is a marketing tool used by new coins. The team behind a new coin or token will give users the ability to receive the asset for free, typically in exchange for a small task, such as following the company on social media or providing an email address. Airdrops are great for projects, since many new customers get excited about the coin and want to see it rise in value. They're also great for users since the coins or tokens are given out for free. However, airdrop scams are common, and many new coins fail, so make sure to do your research to understand which new airdrops are good, and which airdrops aren't. Here are a few sites that provide information on new airdrops:

- aidrops.io
- airdropalert.com
- icomarks.com
- cocoricos.io

Algorithm

Algorithms define rules that software follows. Many forms of algorithms are used across the internet, such as those used by social media services to decide which content gets how much exposure. Blockchains and cryptocurrencies use algorithms to perform a variety of tasks.

Altcoin

Bitcoin was the first cryptocurrency, as well as the coin that popularized the industry. As a result, Bitcoin belongs to its own category, while all other coins are referred to as altcoins.

Angel Investors

Angel investors are wealthy individuals looking to invest in and provide funding to start-ups and entrepreneurs.

Arbitrage

Multiple exchanges trade the same cryptocurrencies at any given time and sometimes small differences in price occur between these exchanges. Arbitrate traders try to profit on this small margin by buying and selling the difference.

Atomic Swap

An atomic swap is a smart contract technology that allows users to exchange two different coins for each other without a third-party intermediary, usually an exchange, and without needing to buy or sell. Centralized exchanges such as

Coinbase cannot perform atomic swaps. Instead, decentralized exchanges allow for atomic swaps and give full control to users.

Bag

A bag refers to one's position. For example, if you own a sizable quantity in a coin, you own a bag of them.

Bag Holder

A bag holder is a trader who has a position in a worthless coin. Bag holders often hold out hope on worthless positions.

Bear / Bearish / Bull / Bullish

To be a bear means that you think the price of a coin, token, or the value of the market as a whole is going to go down. Investors who think like this are considered "bearish" on the given asset. The opposite is to be bullish: a person who thinks a rise in value will occur is bullish on that asset.

Bear Flag / Bull Flag

A bear flag is an indication on a coin or token chart that a downtrend is likely. A bull flag is the opposite of a bear flag and signals an uptrend.

Bear Trap / Bull Trap

A bear trap refers to false downtrend signal. This is referred to as a "trap" because traders who take a bear trap as an indication of a downtrend may short the coin or token; hence losing money when the price actually increases. A bull trap is the opposite of a bear trap. Bear traps can sometimes be manipulated into existence. In such a case, a group of traders aims to quickly crash a cryptocurrency and then make a quick profit on the rebound. Those involved must own a sizable position of a given coin. Then, they all sell their positions at the same time, which bluffs the market into thinking a crash is occurring, which triggers more selling, causing stop losses to be hit and more selling to occur, leading to an even sharper fall. Those who set the trap proceed to buy back into their position at the lower price. Once the price rebounds, they make a profit.

BearWhale

A whale is someone who holds a very large position in a coin. A bear is someone who has a negative opinion of where a coin or token is headed. So, a BearWhale is someone who holds a very large position in an asset yet thinks that asset will decline in price. BearWhales, obviously, are not very common.

Bit

A microbitcoin, or bit, refers to one-millionth of a bitcoin.

Bitcoin

Bitcoin was the first cryptocurrency. It was created in 2008 by an individual or group of individuals operating under the name of Satoshi Nakamoto.

Bitcoin ATM

A Bitcoin ATM is an ATM from which bitcoin, and perhaps other cryptocurrencies, can be bought or sold. Bitcoin ATMs are gaining in popularity in parallel to the growth of Bitcoin and the wider cryptocurrency industry.

Bitcoin Improvement Proposal (BIP)

The Bitcoin Improvement Proposal process is the standard process through which Bitcoin updates can be suggested and the network improved. With the BIP, users can propose changes, and if these changes are universally agreed upon, the entire Bitcoin network may incorporate them.

Black Swan Event

A black swan event is an entirely unexpected occurrence. For example, if you invest in a coin and the next week news comes out that the project received funding from a group of esteemed angel investors, that coin may unexpectedly rise in value.

Block

A block, contextually used as part of a blockchain, is a data structure that contains information about transactions. Such data usually includes the amount and time of the transaction, as well as the addresses involved.

Block Explorer

Block explorers are online services that track blockchain transactions and provide access to blockchain public ledgers. Etherscan is a popular block explorer.

Block Height

The block height is the number of blocks in a blockchain. Height 0 is the first block, (referred to as the genesis block) height 1 is the second block, and so on. Currently, the block height of Bitcoin is past half a million. The "block generation time" of Bitcoin is currently around 10 minutes, meaning that one new block is added to the Bitcoin blockchain approximately every 10 minutes.

Block Reward

A block reward refers to the number of coins a miner can earn per each successfully mined and validated block.

Blockchain

A blockchain is a type of database that organizes each list of transactions (referred to as blocks) into chains, hence the name. Blockchain networks use DLT (decentralized ledger technology) and peer-to-peer networks to create decentralized, anonymous, and secure networks.

Bounty

A bounty is a reward offered to complete a task. Many crypto start-ups will offer bounties (the bounties are typically the coin or token at hand) in reward for users spreading the word about their project.

Brain Wallet

A brain wallet is an account created from a password or passphrase of your choosing. However, since humans are not very good at generating and remembering long and complex passwords, brain wallets are somewhat insecure and can be hacked through brute force attacks. Rather, it's much better to store passwords, so to speak, off-brain, and in secure physical or online locations.

Brute Force Attack (BFA)

A brute force attack is an attack that simply "brute forces" information, such as a password, by simply trying as many combinations as possible. Advanced brute force attacks can generate millions of combinations per second. BFAs are the reason you can only try to enter a password a few times before you get locked out.

Bubble

A bubble in crypto and all investments refers to a time in which most of the market is going up, usually at an unsustainable rate. Often, bubbles will pop and trigger a large crash. For this reason, being in a bubble, whether referring to the or a specific coin or token, is not such a good thing.

Bug Bounty

A bug bounty is a reward given to a person or people who find errors or vulnerabilities in a computer program or the wider system. Large companies often offer millions of dollars in bug bounty rewards, as identifying vulnerabilities could prevent hacks.

Burning

The act of burning a cryptocurrency involves sending coins to an inaccessible account (called an eater address), which reduces the effective coin supply. Burning manages inflation and increases value through scarcity.

Buy Wall

A buy wall occurs when a large limit order is placed to buy a cryptocurrency at a certain value. This "buy wall" can prevent the asset from falling below that value, since demand at that price far exceeds supply.

Cash

In the world of crypto and investments, cash does not mean keeping literal cash, but rather money that is not invested and rather being held in a digital account.

Central Ledger

A central ledger is a ledger controlled by a single entity. Banks use central ledgers.

Chain Linking

Chain linking refers to the process of one cryptocurrency being transferred to another. Since each currency has its own blockchain and the transaction must be recorded on both blockchains, the chains of data on each blockchain is linked to complete the transaction.

Chain Split

A chain split is the same as a fork.

Cipher

A cipher is the name given to any algorithm, offline or online, that encrypts and decrypts information.

Circulating Supply

The circulating supply is the total number of coins in a cryptocurrency that are available to be publicly traded. The circulating supply can be "burned" or increased through mining rewards until the max supply is reached.

Cold Storage

Cold storage refers to the offline storage of passwords or crypto holdings.

Confirmed

Transaction confirmation refers to a transaction being confirmed, which means multiple peers in the network have validated the given transaction. Once a transaction has been confirmed, it is permanently stored and viewable in the public ledger.

Confluence / Confluence Trading

Confluence occurs when multiple strategies and indicators are combined into one strategy. Confluence trading is an extension of this; it refers to a trader who utilizes confluence in their trading strategy.

Consensus

When a transaction is made on a blockchain network, many different nodes in the blockchain must validate it and reach a consensus on whether or not the transaction is valid. Consensus simply means majority opinion.

Consortium Blockchain

A consortium blockchain is a blockchain network that is privately owned by multiple parties.

Corporate Cryptocurrency

A corporate cryptocurrency is a cryptocurrency created or managed by a corporate entity.

Correction

A correction is a price movement downward after a quick jump or a peak in price. For example, a move from $10 to $25 may result in a correction to $20, at which price support is found.

Cryptographic Hash Function

A cryptographic hash function is a certain process that happens within nodes. Each node will convert a transaction (or any other input) into an encrypted string made up of letters and numbers that registers the place of the transaction in the blockchain.

dApp / DAOs

dApp is short for "decentralized application." Basically, any app that runs on a blockchain (or any other peer-to-peer network) and does not have a centralized owner is considered a dApp. DAO is shorthand for decentralization autonomous organization and refers to any organization that is transparent, owned by a network of distributed participants, and run by programmed rules as opposed to a centralized structure.

Dead Cat Bounce

A dead cat bounce is a term that refers to a brief price recovery before a large crash.

Decryption / Encryption

Encryption is the process of converting plain text into coded information through the use of a cipher. The opposite is decryption, which converts coded information into plain text. Decryption in crypto involves turning encrypted data back into plain text.

Deep Web / Dark Web

The deep web consists of parts of the web that are not indexed by search engines and therefore inaccessible by search. This is not innately nefarious and is simply how web pages are kept private. The dark web is also not necessarily used for nefarious purposes and consists of hidden websites which are only accessible through specialized web browsers such as Tor.

Delisting

Sometimes, a coin or token get removed from an exchange. Either the exchange or the team behind the project may choose to delist the asset. The most well-known occurrence of widespread delistings followed an SEC investigation into Ripple (XRP).

Depth Chart

Depth charts graph buy and sell requests. The depth chart shows the crossover point at which transactions are quickly completed, which is the market price.

Derivative

A cryptocurrency derivative is a financial product that derives value from an underlying asset, such as a coin or token, and allows investors to hedge their bets and mitigate loss.

Difficulty

Difficulty, in the crypto space, refers to the cost of mining. The difficulty can change from moment-to-moment based on demand and supply.

Digital Commodity

A digital commodity is a digital asset that holds value. Digital commodities do not have to be digital currencies. NFTs, digital art, and anything else that holds value and exists online are digital commodities.

Digital Currency

Digital currencies lie within the realm of digital commodities. Instead of referring to all digital assets, digital currencies refer to all currencies that operate solely online and do not have a physical form.

Digital Signature

Your digital signature is used to confirm that online documents come from you. This is not equivalent to a physical signature. Instead, digital signatures are code generated by an algorithm.

Distributed Ledger

A distributed ledger is a ledger that is stored in many different locations so that transactions can be validated by multiple parties. Blockchain networks use distributed ledgers.

Dolphin / Whale

Crypto holders are classified through size of holdings. Those with extremely large holdings are called whales, while those with moderately sized holdings are called dolphins.

Double Spend

Double spending occurs when a cryptocurrency is sent to two different wallets at the same time. Blockchains leverage various consensus models to stop double spending and such attacks are practically impossible to carry out on most major cryptocurrencies.

Dump

To dump, or dumping, refers to selling a large amount of cryptocurrency or to a large amount of a coin or token being sold. For example, "that coin is dumping," and "I'm dumping this coin."

Dust Transaction

A dust transaction is an extremely small transaction. Sometimes, attackers flood networks with dust transactions to track down the transactional activity of certain wallets. Through a combined analysis of address activity, the attackers may be able to deanonymize the person or company behind the wallet.

Efficient Market Hypothesis (EMH) / Adaptive Market Hypothesis (AMH)

The EMH is an economic theory stating that the price of publicly tradeable assets reflects all publicly available information. Given the rise of coin and tokens that move in value simply due to hype, pump-and-dump schemes, and all the other ways in which an asset may uptrend without underlying fundamental reasons, the Efficient Market Hypothesis is not reflective of the current cryptocurrency market.

The Adaptive Market Hypothesis, AMH, proposes that financial markets are governed by the laws of biology. Those laws notably including the following: people act in their own self-interest, people make mistakes, and people adapt and select based on past actions. The AMH is more present in current crypto markets than the EMH.

Encryption

Encryption is the process of converting plain text into coded information through the use of a cipher. The opposite is decryption which converts coded information into plain text.

End Users

End users are the people who use the final version of a product (e.g., at the end of the creation process). So, developers and beta testers are not end users, while a consumer who buys a product from a major retailer is an end user.

ERC-20 / ERC-20 Standard

An ERC-20 is one of the many Ethereum token types. Remember, a token is a token because it is built upon another blockchain, while coins are built on their own blockchains. ERC-20 is significant in the world of Ethereum tokens because it is used to define the rules by which all tokens on the Ethereum blockchain function. It can be likened to a security guard; it requires and ensures that all tokens in its vicinity follow that set of rules. The ERC-20 "standard" is the combined list of all the rules. Tokens using the ERC-20 standard can transact between each other and exchange in a more efficient manner.

Escrow

Escrow refers to a third party that holds funds during a transaction. This third party should be unbiased and makes sure that both parties stick to an agreed-upon deal.

Ether

Ether is the native cryptocurrency of the Ethereum blockchain. Its ticker symbol is ETH, and to use any currency on the Ethereum blockchain you must pay fees in Ether.

Exchange

A cryptocurrency exchange is a marketplace in which cryptocurrencies are traded. Exchanges must be combined with wallets. In wallets, coins can be held through addresses. Exchanges act as an easy intermediary to help users transact.

Faucet

A faucet is a website that offers free cryptocurrency in exchange for information, such as an IP address. Faucets are scams and differ from Airdrops. Never give personal information away online, especially in return for promised "free" cryptocurrency.

Fiat

Fiat refers to governmental currencies, such as the US dollar and Euros.

Fill or Kill Order (FOK)

A fill or kill order is an order that must be executed upon immediately. If this doesn't happen, the trade will be canceled. FOKs are used to ensure that transactions involving large positions will be completed in a very short period.

FinTech

Fintech is short for financial technology. Fintech consists of any technology that supports and/or enables financial services. Cryptocurrencies are fintechs, as well as companies such as GoFundMe and PayPal.

Flippening / Flappening

The "flippening" is used to describe the hypothetical moment when, if at all, Etherium (ETH) passed bitcoin (BTC) in market cap. The "flappening" was the moment when Litecoin (LTC) passed Bitcoin Cash (BCH) in market cap. The flappening happened in 2018, while the flippening has yet to occur, and, based purely on market cap, is presently unlikely.

Fork

A fork is the occurrence of a new blockchain being created from another blockchain. For example, Bitcoin Cash once forked off from Bitcoin. Forks occur when algorithms have a disagreement and split into two different versions. Two kinds of forks exist: A hard fork and a soft fork. A hard fork in a blockchain is a fork that occurs when all the nodes in the network upgrade to a newer version of the blockchain and leave the old version behind. Two paths are then created: the new version and the old version. A soft fork contrasts this by rendering the old network invalid; this results in just one blockchain, not the two that comes as a result of a hard fork.

Frictionless

A frictionless market is an ideal trading environment in which no costs or restraints exist on transactions. Frictionless markets are only theoretical.

Full Node

A full node is a node that downloads and contains the entire history of a blockchain in order to completely enforce its rules without requiring the assistance of other nodes.

Fundamental Analysis

Fundamental analysis is analysis of a coin or token through its fundamental metrics. Fundamental metrics looks at economic and financial activity to determine value.

Futures Contract

A futures contract is a pre-approved contract between two entities to fulfill a transaction at a certain price or date.

Gas

Gas refers to the fee required to complete transactions on the Ethereum blockchain. Gas fees are given to the miners, who validate blocks and ensure secure networks.

Gas Limit

When a transaction is made on a network, a user can set the limit they wish to pay in gas fees. Manually setting higher gas limits will result in transactions being fulfilled quicker, since the reward is higher. Gas price is typically set automatically to the going market rate.

Genesis Block

The genesis block is the first block in a blockchain.

GitHub

GitHub is a collaboration platform for coders and software developers, in which open-source code can be shared, implemented, and improved.

Golden Cross

A golden cross is a chart pattern involving a short-term moving average (say, a 10-day MA) crossing above a long-term moving average (perhaps a 50-day MA). Golden crosses are bullish indicators.

Group Mining / Mining Pools

Group mining refers to groups of people or entities who combine their computational power in order to mine together and split the rewards. Group mining is synonymous with mining pools.

Gwei

Gwei is the denomination (the price-per-unit) used in defining the cost of Etherium gas. You can think of Gwei and Etherium as similar to the penny versus the dollar. 1 ETH equals one billion Gwei. Gwei is used instead of Etherium because seeing that gas fees are 1 Gwei is easier than seeing the fee as 0.0000000001 Ether. That said, gas fees are quite high as of 2022, and for this reason resorting to Ether denominations is currently more applicable, though this will not be the case forever.

Halving

Halving is the process by which the reward for mining Bitcoin is cut in half. Bitcoin halving happens every 210,000 blocks, which roughly equates to every 4 years. Halving will happen until the maximum supply of bitcoin has been reached and all 21 million coins are put into circulation.

Hard Cap

A hard cap is the maximum amount of money the creator(s) of a coin can raise during an ICO (initial coin offering).

Hard Fork / Soft Fork

A fork is the occurrence of a new blockchain being created from another blockchain. For example, Bitcoin Cash once forked off from Bitcoin. Forks occur when algorithms have a disagreement and split into two different versions. Two kinds of forks exist: A hard fork and a soft fork. A hard fork in a blockchain is a fork that occurs when all the nodes in the network upgrade to a newer version of the blockchain and leave the old version behind. Two paths are then created: the new version and the old version. A soft fork contrasts this by rendering the old network invalid; this results in just one blockchain, not the two that comes as a result of a hard fork.

Hash / Hash Rate

A hash is a function that converts one value into another; a hash in the crypto world converts an input of letters and numbers (a string) into an encrypted output of a fixed size. Basically, hashes help with encryption. "Solving" each hash requires working backwards to solve an extremely complex mathematical problem. The measure by which a computer is judged in terms of its ability to hash is called a hash rate. Put simply, the hash rate is the speed at which a node can perform hashing, and hashing is important in cryptography.

Hexadecimal (Hex)

The hexadecimal system is a numbering system used to reduce the amount of work needed by computers. The Hex system is based off 16 symbols—0 through 9 and A through F.

Hot Wallet / Cold Wallet

A hot wallet refers to a cryptocurrency wallet that is connected to the internet. The opposite, cold storage, refers to a wallet that is not connected to the internet. Hot wallets allow for the owner of an account to easily send and receive tokens; however, cold storage is more secure than hot storage.

Iceberg Order

An iceberg order executes a large order through many smaller orders. This allows the orders and the buyer or seller to stay somewhat discreet, as iceberg orders are used by those wanting to keep large transactions under the radar.

Index Fund / Basket Trading

A crypto index fund consists of many cryptocurrencies; all of which can be invested in at once. For example, an index fund made up of bitcoin (BTC) and Etherium (ETH) can allow you to invest in both securities without having to buy both separately. Index funds and basket trading also allow users to diversify risk; instead of investing in just one volatile asset; you spread your risk over many. An index fund can contain dozens, or hundreds, of securities.

Inflation

Inflation is the price increase of goods and services as a result of more money being printed. For example, say one million dollars exist and bread costs $1. If another one million dollars are printed, the price of bread should logically double to $2, since two times as much money exists. Since more dollars exist, each individual dollar is worth less, making goods and services cost more, and this is the core concept of inflation.

Initial Coin Offering (ICO)

In order to raise funds and awareness, the creators of a cryptocurrency often put an initial portion of their coin supply up for purchase.

Initial Exchange Offering (IEO)

An IEO is similar to an ICO. Both are initial offerings of coins or tokens used solely within the crypto space. IEOs are coming into fashion as the improved version of ICOs because IEOs allow online crypto trading platforms to directly make the asset tradeable. Basically, IEOs require less effort to invest in and streamline the trading process of an initial offering.

Initial Public Offering (IPO) / Direct Listing

An IPO is the process by which a company can become publicly traded on a stock exchange. Crypto coins or tokens can have an IEO or an ICO, but not an IPO. However, some centralized crypto companies grow large enough that they have an IPO. A company can also become listed on a stock exchange through a direct listing, and both methods have a common end result. Coinbase was the largest IPO of a crypto exchange.

Instamine

An instamine occurs when coins are created as a single large batch, as opposed to slowly creating new coins through mining. Instamines are not very common and more susceptible to fraud.

Keys

A key is a random string of characters used by algorithms to encrypt data. Two keys are used for cryptocurrency: a public key and a private key. Both are important to understand and are defined in depth throughout other sections.

Key Pairs

A key pair is a combination of a public and private key. All wallets have a unique key pair attached to them.

Know Your Customer (KYC)

Know your customer guidelines requires that financeal professionals get information about the identity of their customers. KYC procedures fit within the wider AML (anti-money laundering) policies.

Latency

Latency is the delay between when a transaction is submitted and when the network recognizes the transaction. Latency may be thought of as lag.

Layer 2

Layer 2 is a secondary framework or protocol built on top of a blockchain system. Most layer 2 protocols are designed to boost blockchain scalability.

Ledger

A blockchain ledger stores data about all financial transactions made on a given blockchain. Cryptocurrencies use public ledgers, meaning all transactions performed with that cryptocurrency are publicly viewable through a ledger. Refer to the Blockchain section for more information on public ledgers.

Leverage

Investors can "leverage" their money by taking on debt. Say you have $1,000 and you take on 5x leverage; you're now able to invest $5,000 worth of funds. By that same function, 10x leverage is $10,000 and 100x is $100,000. Like margin trading, leverage allows amplified profits by, in a sense, renting money and pocketing the extra profit. However, leverage trading is very risky; unless you are an experienced trader and financial stable, leverage trading is not recommended.

Lightning Network

A lightning network is a type of secondary layer on top of a blockchain system. Lightening networks enable faster transactions.

Limit Order / Buy / Sell

When you execute a trade, you may choose to have that trade executed in several different manners. One of such manners is through a market order, which executes orders immediately at the best market price available. The popular alternative is a limit order, which lets the buyer or seller choose the price at which they want to buy or sell.

For example, say a coin is trading at $200. If you choose to buy one coin with a market order, that order will immediately execute, perhaps at $200, or perhaps at $199 or $201. If you place a limit order, you choose the price you want to purchase the one coin at. Maybe this coin is volatile, so you decide to place a limit buy order at $197 in the hopes of the price spiking down to that level at some point throughout the day and before recovering and continuing in an uptrend. In this case, the order would only execute when the price of the coin hits $197 or below.

Generally, limit orders are good for catching the price slightly below the market value at the time of the order, though limit orders can be set at any price for various other reasons. If you're placing and order and don't really care whether your purchase point is 2% lower or not (or whatever it may be), you can just set a market order and purchase the security instantly. Perhaps you are worried that the price will keep increasing and won't dip. In that case, a market order is preferable over a limit order.

Listing

A listing is the addition of a coin or token to an exchange.

Liquidity

Liquidity is how easily an asset can be bought or sold. For example, stocks and cryptocurrency are extremely liquid, since they can be bought or sold at a moment's notice. However, assets such as real estate and priceless art may be less liquid, since to sell them requires significant time, effort, and money.

Locktime

Some transactions come with a rule that delays the time at which the transaction can be validated and confirmed. This is called locktime. The vast majority of blockchains do not involve locktime.

Long / Short [position]

Taking a long position means that an investor intends to hold an asset for the long-term; this generally means at least a few months. A short position is the opposite; the trader intends to get in and get out in a relatively short time period, whether minutes, hours, or days.

Mainnet

The mainnet constitutes the main network of a blockchain. Transactions take place on the distributed ledger, not the mainnet.

Mainnet Swap

A mainnet swap occurs when a coin moves from one mainnet to another.

Maker

Not to be confused with the Maker coin. A user becomes a maker when they place an order, and that order is not immediately fulfilled. That order becomes open and is placed in the order book until fulfilled.

Margin Trading

Margin trading is a popular strategy whereby traders borrow funds to place trades. For example, someone with $10,000 may trade with 5x margin, giving them $50,000 of capital. If the trade works out, they repay the $50,000 (typically with interest or some type of fee) and keep the extra profit. Margin trading should only be performed by experienced investors—if trades go bad, many traders end up with more debt than money. So, the rewards are massive, but the risks are just as exceptional.

Market Capitalization (Market Cap)

The market cap of a coin is the total trading value. This can easily be calculated by multiplying the total supply of a coin by the price of the coin. For example, a cryptocurrency trading at $5 with a supply of 1 million units has a market cap of $5 million.

Market Momentum

The momentum of a market is the ability of that market to maintain periods of growth or shrinkage. A market that has been in the green for six months has strong momentum, while the same can be said if that market plunges into bear territory and stays in the red for significant periods of time.

Market Order

A market order is one several types of orders that can be placed to execute a trade. Market orders are immediately filled at the best market price available. The opposite, limit orders, allows the buyer to choose the price they want their trade to be fulfilled at. While market orders may result in the purchase point being slightly higher than a smart limit order, they allow for quicker entry.

Masternode

Masternodes have more roles than a regular node in a blockchain network, such as enabling specific services.

Maximum Supply

The maximum supply of a coin or token, as the name suggests, is simply the total amount of coins that can possibly be created for a given cryptocurrency. When a coin or token is created, the maximum supply is built into the algorithm. Once the maximum supply has been reached and all the coins or toes have been mined, no more can exist. For example, Bitcoin has a maximum supply of 21 million, which will be reached around the year 2140.

Mempool

A mempool is a node's holding area for pending (unvalidated) transactions.

Merged Mining

Merged mining is the act of mining multiple cryptocurrencies at the same time.

Merkle Tree

A merkle tree is a way in which data is structured. The Merkle structure gives off the visual appearance of being a tree, and Merkle Trees are synonymous with Hash Trees.

Mining

Mining is the process by which blocks are added to a blockchain through working backward to solve mathematical problems. Solving these problems takes an extremely large amount of computational power. To make up for this cost and incentivize activity, rewards are provided to those who do the work. The people or organizations who use their computational power to mine are the miners.

Mining Contract

A mining contract is a contract that involves one party loaning (basically, renting) their computational power (hashing power) to another party. The purchaser pays an upfront fee in return for the rewards generated by the rented hashing power.

Mining Farm

A mining farm is a collection of multiple miners, typically a large group, who manage a large data center or warehouse devoted specifically to mining cryptocurrencies. Dave Carlson manages one of the largest mining farms in the world. His monthly expenses are more than $1 million worth of electricity, and although specific numbers are not public, it's estimated that Carlson at some point mined up to 200 bitcoins per day. Please keep in mind that mining farms are illegal in many places around the world.

Mining Pool

A mining pool is a group of miners who combine their computational power to earn rewards faster. The rewards are then split across the group relative to the amount of power contributed. A mining pool is better than individual mining for those with relatively little computing power because rewards are distributed more often. So, as opposed to a small miner earning one big dump of $1000 worth of cryptocurrency every two years, that miner may join a pool and earn $1.37 per day.

Mnemonic Phrase

A mnemonic phrase is synonymous to a seed phrase. Both terms describe 12-to-24word sequences that identify and represent a wallet. It is like a backup password, and anyone with it has complete access to a cryptocurrency wallet. Make sure to securely store your mnemonic phrases.

Moon

Phrases such as "to the moon" and "it's going to the moon" refer to a cryptocurrency going up in value, typically by an extreme amount.

Multipool Mining

Multipool mining is the event of miners moving from one cryptocurrency to another based upon the rewards offered. Pools offering multipool mining automatically switch between coins. Rewards are then distributed to the miners who contributed computational power.

Multisignature

Some wallets require multiple parties to authorize and validate transactions before such transactions are added to the public ledger, hence requiring a multisignature. Multisignature addresses and multisignature wallets require multisignatures.

Network

A network, at its core, is an interconnected system. The system within a cryptocurrency network is made up of many nodes that assist the blockchain in a variety of tasks.

Node

A node is a computer connected to a blockchain's network that assists the blockchain in writing and validating blocks. Some nodes download an entire history of their blockchain; these are called masternodes and perform more tasks than regular nodes. Additionally, nodes are not locked into a specific network. Rather, most nodes can switch to different blockchains practically at will, as is the case with multipool mining.

Nonce

A nonce is an arbitrary number used just once to verify a cryptographic transaction. The nonce is the number crypto miners are looking to find by solving mathematical equations.

Oracles

Smart contracts within a blockchain can only be externally reached through oracle programs. Oracles send data to and from smart contracts and external sources as required; you may think of them as performing the same tasks as messenger RNAs in the human body.

Order Book

An order book is a list of open buy and sell orders for an asset on an exchange. Any order that has not been filled is an open order in an order book.

Oversold / Overbought

An oversold cryptocurrency experienced much more selling pressure than buying pressure. As a result, it has been sold down to a price that is considered below its fundamental true value. Therefore, being oversold generally means that the security should rebound at least to its true value.

Overbought is the opposite and occurs when a coin or token has been bought up to what may be considered an unjustifiably high price. Usually, if someone believes that a coin or token is oversold, they believe that it will go up, while if they believe it is overbought, they believe it will go down.

Paper Wallet

A paper wallet is a method of storing cryptocurrency keys and seed phrases, which, as the name implies, is simply printed out on paper.

Peer-to-Peer (P2P) / P2P Networks

A peer-to-peer network involves many computers working with each other to complete tasks. Peer-to-peer networks do not require a central authority and are an integral part of blockchain networks.

Pegged Currency

A pegged currency, like a stablecoin, is a currency designed to remain at the same price as that of a designated asset, typically a bank-issued currency. USDT and DAI are two popular stablecoins pegged to the US dollar, meaning that 1 DAI and 1 USDT will be equivalent in perpetuity to the US dollar.

Permissioned Blockchain / Ledgers

Most crypto blockchain networks are public; this means that anyone can mine the blockchain and assist in adding blocks. An alternative to this system is a permissioned blockchain, which includes a corresponding permissioned ledger. Within permissioned networks, an access control layer is built on top of the normal blockchain that controls, in a nutshell, who can do what.

Pre-Sale

A pre-sale occurs in a start-up just about to launch; invited investors are able to purchase the token before it becomes publicly available.

Price Action

Price actions are simply the price moments of an asset over time.

Private Key / Public Key

Cryptocurrency users will utilize two keys: a public key and a private key. Both keys are strings of letters and numbers. Once a user initiates their first transaction, a pair of both a public key and a private key is created. The public key is used to receive cryptocurrencies, while the private key allows the user to carry out transactions from their account. Both keys are stored in a crypto wallet.

Private Sale

A private sale is an early-stage investment round that is not open to the general public, and instead is typically only open to investors with a significant amount of funds.

Proof of Authority (PoA)

A PoA algorithm gives a small number of users the authority to validate transactions. This saves computational power and creates a streamlined validation process.

Proof of Burn (PoB)

A PoW algorithm is used to confirm transactions and create new blocks on a given blockchain. "Proof of work" literally means that work, through mathematical computation, is required to create blocks. The people who own the computers that perform the computations are miners.

Proof of Stake (PoS)

PoS algorithms allow users to mine and validate transactions based on the size of their holdings. So, with PoS, the more you own, the more you can mine. Established PoS staking rewards usually vary between 5% and 20%.

Proof of Work (PoW)

A PoW algorithm is used to confirm transactions and create new blocks on a given blockchain. "Proof of work" literally means that work, through mathematical computation, is required to create blocks. The people who own the computers that perform the computations are miners.

Protocol

A protocol is a system or procedure that controls how something should be done. Within cryptocurrency, protocols are governing layers of code. For example, a security protocol determines how security should be carried out, a blockchain protocol governs how blockchain acts and operates, and a Bitcoin protocol controls how the Bitcoin network functions.

Public Blockchain / Private Blockchain

A public blockchain is an open network that allows any computer to participate in verifying transactions. On the other hand, private blockchains regulate who has access to the network and who may participate in the network.

Pump / Dump

A pump is a rapid upward price movement in a coin or token. A dump is a rapid downward price movement in a coin or token. "To the moon" refers to a massive pump.

Pump and Dump

A pump and dump is a scheme carried out by a large investor or, more typically, a group of large investors. In a pump and dump, the starting group will buy a significant amount of a coin or token. Other investors see the strong upwards pressure and buy in. Then, once the price has been significantly inflated, the original investors dump their shares and take profits. This practice is looked down upon since it is manipulative and causes most investors involved to lose money.

QR Code

QR stands for "quick response." Within the cryptocurrency ecosystem, QR codes are commonly used to make wallet addresses scannable, as opposed to necessitating manual transcription of the entire key.

Race Attack

A race attack, which is a subset of double-spend attacks, occurs when two transactions are created with the same funds at the same time, with the goal of spending the money twice and doubling the initial amount.

Rank / Ranking

Cryptocurrencies are ranked by market cap. Within the ranking system, which may be thought of like a scoreboard, being in the top 10 is equivalent to a badge of honor.

You'll often hear people say, "I think this could be a top 10 coin," and similar statements. Bitcoin has held the top spot since inception and is likely to hold that spot for at least another few years. Check out the coin rankings for yourself at any of the following sites:

- ◆ coinmarketcap.com
- ◆ coingecko.com
- ◆ cryptoslate.com

Refund Address

A refund address is a wallet address that serves as a backup in case a transaction fails. If such an event occurs, a chargeback is given to the specified refund address. Refund addresses are not common, and in most cases, transactions send to the wrong address are non-returnable.

REKT

Rekt, as you may have guessed, is slang for "wrecked" and is used to describe a bad trade or large loss.

Resistance

Resistance is a price that an asset struggles to break through in an upwards manner. Sometimes, levels of resistance can be phycological. For example, bitcoin may hit resistance at $100,000, since many people place sell orders at the nice, round, and pleasant-sounding price of $100,000. When a resistance level is broken through, price can quickly climb. In a situation where bitcoin breaks past $100,000 following a period of strong resistance, the price may quickly climb to $105,000. Support is the opposite of resistance.

Ring Signature

A ring signature is a digital signature (it can also be thought of as an encryption process) that allows both the giver and the receiver to remain anonymous by giving nodes within the network the power to approve transactions without identifying which node requested the transaction, hence removing any digital trail between the two nodes and keeping the keys and the identity of the sender and receiver private.

Roadmap

A roadmap is a plan that an organization publishes regarding its long-term goals and important benchmarks.

Satoshi Nakamoto

Satoshi Nakamoto is the individual or possibly the group of individuals who created Bitcoin. Not much is known about this mysterious figure, and his anonymity has spawned countless conspiracy theories. While Nakamoto lists himself as a 45-yearold male from Japan on an official peer-to-peer foundations website, he uses British idioms in his emails. Additionally, the timestamps of his works better align with someone based in the US or the UK. Nakamoto currently holds a fortune worth more than $50 billion through holdings of 1.1 million bitcoins.

Scrypt

A scrypt is an algorithm that encrypts data (notably keys) in a way that makes brute forcing the information very difficult. Scrypt-encrypted keys requires an enormous amount of computational power to hack (technically, to hash), which raises the barrier of entry for potential attackers.

SEC

The Securities and Exchange Commission, (SEC) is a governmental agency responsible for regulating security markets, such as the stock market.

Security Audit

A security audit is an analysis of how safe a system is against attacks or technical failures. Companies often perform security audits in order to improve their security measures.

Seed / Seed Phrase

A seed phrase is interchangeable with a mnemonic phrase. Seed phrases are 12-to24-word sequences that identify and represent a wallet. With it, you can never lose access to a connected account. If you forget it, there's no way to reset it or get it back. Anyone who has your seed phrase has full access to the connected wallet and cryptocurrency holdings.

Segregated Witness (SegWit)

A SegWit process allows more transactions to fit within one block by separating transaction signatures.

Selfish Mining

Selfish mining is a group mining strategy in which the miners strategically withhold blocks to increase profit.

Sell Wall

A sell wall is a very large sell order at a limit price. Sell walls drive prices downwards. The opposite is a buy wall, which can stop a coin or token from falling beneath a certain price.

Sentiment

Sentiment describes the attitude of a person or group of people towards something, perhaps a company, person, market, or asset. In a nutshell, sentiment is emotion.

Sharding

Sharding is the process of splitting up a blockchain network into smaller portions. In this way, transactions can be recorded and stored in only one shard, as opposed to going through every single node in the network. Sharding enables blockchain scalability and speed.

Slippage

Slippage can occur when a trade is placed through a market order. Market orders try to execute at the best possible price, but sometimes a notable difference occurs between the expected price and actual price. For example, say you want to buy 20 BNB for $1000, but you place a market order and only end up getting 9 BNB for $1000. Slippage is rarely this drastic, but regularly varies in the 1-3% range. When placing large orders, it is usually better to place a limit order as opposed to a market order. This eliminates the danger of slippage.

Smart Contracts

Smart contracts are an essential part of the cryptocurrency world. A smart contract is a self-executing contract run on code. The terms of the contract, as well as the execution, is directly written into the smart contract and therefore removes the issue of trust for all parties in the transaction. Transactions issued with smart contracts are irreversible and untraceable. These contracts can be used not just for managing cryptocurrency transactions, but also in voting systems, various financial services, information storage, and throughout a multitude of other industries.

Solidity

Solidity is a programming language, just like Python or Java, which is used to write and develop Ethereum smart contracts.

Source Code

Source code defines how software functions based on programmed rules.

Stablecoin

A stablecoin, similar to a pegged currency, is a coin or token that is designed to remain at the same price as a designated asset, typically a government-issued currency. For example, USDT and DAI are two popular stablecoins pegged to the US dollar, meaning that 1 USDT and 1 DAI will each exist in perpetuity as equal to 1 US dollar. Stablecoins experience practically no volatility, typically provide a few percent interest (APY) on holdings per year and are a good place to store money in the crypto ecosystem.

Staking Pool

A staking pool is a group in which stakeholders combine their computational staking power in order to increase their ability to successfully validate a new block. A block reward is earned each time a block is validated, and rewards are thereafter distributed in accordance with each member's contribution.

Stale Block / Orphan Block

Due to the geographical separation of nodes in a blockchain network, multiple chains, each with multiple blocks, can simultaneously exist. Sometimes, two miners, each in different geographical locations, may hash (solve) a block at practically the same time. Both blocks are valid, and both reach different chains. One block will be "chosen" as the truth, with the choice being decided by the length of the chain. The block that did not get accepted is known as a stale or orphan block.

State-Based / State-Sponsored Cryptocurrency

State-based cryptocurrencies are coins or tokens that are sponsored, started, or managed by a central government. As of 2021, China is developing a state-run yen.

Store of Value

Assets that are good stores of value can be acquired and exchanged at a later time for a retained degree of purchasing power.

Support

Support is the price at which an asset struggles to break through in a downward manner since many investors are willing to buy the asset at that price and therefore buying pressure far exceeds selling pressure. Often, if a coin hits support levels, it will reverse into an uptrend. Support level bounces are often good short-term buy opportunities, though if support levels are broken through in a downward manner, a sharp and extended fall is likely.

Swing / Swing Traders

A swing is a dramatic reversal in price. Swing traders try to catch and trade upon market and asset-specific swings.

Tank / Tanked

Refers to a massive dive in price, e.g., "it tanked from $20 to $10."

Technical Analysis

Technical analysis looks at technical indicators to predict price movement. Technical analysts use historical data from charts to make their predictions. Refer to the technical analysis section earlier in the book for an extended look into technical analysis and various charting strategies.

Test Net

A testnet is a piece of software used to test cryptocurrencies before public release. While each testnet is identical to the real software used by the cryptocurrency, the cryptocurrency it trades with is worthless.

Ticker / Ticker Symbol

A ticker is a sequence of letters that identifies a specific coin or token. All stocks, as well as cryptocurrencies, have ticker symbols. For example, bitcoin is symbolized through BTC and Ethereum through ETH.

Timestamp

A timestamp is part of every block within a blockchain. Each timestamp contains the exact moment in which its block was mined and validated. This assists in confirming that blocks do not get tampered with.

Token

While cryptocurrency coins are built upon their own blockchain, cryptocurrency tokens are built upon a non-native blockchain. Many tokens use the Ethereum blockchain, and are thus referred to as tokens, not coins. Token uses are represented under subcategories, the most essential of which are security tokens, platform tokens, utility tokens, and governance tokens. Understanding tokens is an integral part of understanding what exactly you're trading, as well as understanding all uses of digital currencies, and for those reasons we will take a brief look at the token types just mentioned.

- ◆ Security tokens represent legal ownership of an asset, whether digital or physical. The word "security" in security tokens doesn't mean security as in being safe, but rather, "security" refers to any financial instrument that holds value and can be traded. Basically, security tokens represent an investment or asset.

- Utility tokens are built into an existing protocol and can access the services of that protocol. For example, utility tokens are commonly given to investors during an ICO. Then, later on, investors can use their utility tokens as a means of payment on the platform that provided the tokens. The macro definition to keep in mind is that utility tokens can do more than just serve as a means to buy or sell goods and services.

- Governance tokens are used to create and run voting systems for cryptocurrencies that enable functions such as system upgrades.

- Payment (transactional) tokens are used solely to pay for goods and services.

Token Lockup

Token lockups occur when a token or coin is not allowed to be transferred or traded.

Token Sale

A token sale, also commonly referred to as an ICO (initial coin offering), is a limited period of sale in which a set number of new coins are put up for sale in exchange for another cryptocurrency.

Tokenless Ledger

A tokenless ledger is a distributed ledger that does not require a cryptocurrency to function.

Total Supply

Total Supply refers to the total number of coins or tokens that currently exist, whether "burned" or in circulation.

Transaction

A transaction is any exchange between multiple parties. A cryptocurrency transaction involves one party buying a coin or token, and another party selling that coin or token.

Transaction ID (TXID)

TXIDs are transaction identifiers used to reference transactions on a blockchain.

Turing Complete

Turing complete is a term that describes theoretical machines which can solve any computational problem, if given enough time, memory, and proper instruction.

Unconfirmed

Unconfirmed transactions are transactions that have not yet been verified and put onto a blockchain.

Unit of Account

Units of accounts measure asset value. US dollars, as well as other governmental currencies, are units of account because they evaluate things that hold value. For example, if you buy a loaf of bread for $5, the unit of account (the money you used to buy the bread) is being measured against the value of the bread.

Unpermissioned Ledgers

Unpermissioned ledgers are ledgers that have no single owner. The purpose of such a ledger is to allow for all the benefits of decentralization, most notably transparency, efficiency, and security.

Unspent Transaction Output (UTXO)

The UTXO is the amount of a digital currency left over once a transaction has been executed. A crypto wallet contains many UXTOs, all of which represent a certain amount of a coin or token. UXTOs can basically be thought of as smaller bills. When paying with larger bills; they allow you to buy and sell fractions of coins, as opposed to whole quantities of such coins. In this way, UXTOs let you buy one-tenth of a bitcoin as opposed to one.

User Interface (UI)

The UI is the interface through which users interact with software. Each website you visit is displaying its user interface, which lets you interact with the website code.

Vaporware

Vaporware is a coin or token that has been hyped up but has little intrinsic value and is likely to decrease in value.

Virtual Machine

A virtual machine is a computer resource which allows users to run operating systems on computers. These operating systems act as separate computers despite existing purely virtually.

Vladimir Club

A term that describes someone who has acquired 1% of 1% (0.01%) of the maximum supply of a cryptocurrency.

Volatility

Volatility is the size of change in a coin or token, and therefore the ability and likelihood that a price will change rapidly, whether in a positive or negative direction. So, a coin that moves 10% up one day, 27% down the next day, and 22% up the third day is more volatile than a coin that moves up 2%, down 0.5%, and up another 1%. Some coins, specifically stablecoins, have very little volatility, while other coins or tokens, typically those with a relatively small market cap, are extremely volatile and move up and down rapidly.

Wallet

A wallet is the user interface that you use to manage your account(s). Holdings are not actually stored in wallets, which are accessible through a private and public key, but rather on a blockchain. Coinbase wallet and Exodus are common wallets

Weak Hands

Traders who have "weak hands" lack the confidence to hold their assets in the face of volatility and often trade on emotion, as opposed to sticking to their trading plan.

Wei

A wei is the smallest possible denomination of an Ether token.

Whale

A whale is a person or entity with an extremely large position in a coin.

White Paper

A whitepaper is a document that startup companies give to potential investors that reveals information about the company plans.

Whitelist

A whitelist is a list of approved items or participants, such as at an event. The opposite is blacklist, which is a list of banned items or participants.

Wicks / Whiskers / Shadows

Whiskers are the lines extending from the colored bars on candlestick charts and refer to the low-high range of the given asset. Wicks, whiskers, and shadows are synonymous.

Yield Farming

Yield farming puts crypto assets to work in order to generate returns. Yield farming involves users lending funds to others through smart contracts. In return for lended funds, the lender earns rewards in crypto. Yield farming has recently become quite popular.

Zero-confirmation Transaction

A zero-confirmation transaction is a transaction that has not yet been recorded on a blockchain.

Zero-knowledge Proofs

Zero-knowledge proofs verify transactions without revealing information about the transaction, hence keeping the translator's information private while maintaining secure and efficient transaction mediums.

Resources

Resources

- Books
- Exchanges
- Podcasts
- News Services
- Charting Services
- NFT Marketplaces
- YouTube Channels
- Index

Books

- *Mastering Bitcoin* Andreas M. Antonopoulos

- *The Internet of Money* - Andreas M. Antonopoulos

- *The Bitcoin Standard* - Saifedean Ammous

- *The Age of Cryptocurrency* - Paul Vigna

- *Digital Gold* - Nathaniel Popper

- *Bitcoin Billionaires* - Ben Mezrich

- *The Basics of Bitcoins and Blockchains* - Antony Lewis

- *Blockchain Revolution* - Don Tapscott

- *Cryptoassets* - Chris Burniske and Jack Tatar

- *The Age of Cryptocurrency* - Paul Vigna

- For Dummies:
  Cryptocurrency Investing - Kiana Danial
  Bitcoin - Prypto
  Blockchain - Tiana Laurence
  Ethereum - Michael G. Solomon

Exchanges

- Binance - binance.com (binance.us for US residents)

- Coinbase – coinbase.com

- Kraken – kraken.com

- Crypto – crypto.com

- Gemini – gemini.com

- eToro – etoro.com

Podcasts

- What Bitcoin Did by Peter McCormack (Bitcoin)

- Untold Stories (early stories)

- Unchained by Laura Shin (interviews)

- Baselayer by David Nage (discussions)

- The Breakdown by Nathaniel Whittemore (short)

- Crypto Campfire Podcast (relaxed)

- Ivan on Tech (updates)

- HASHR8 by Whit Gibbs (technical)

- Unqualified Opinions by Ryan Selkis (interviews)

News Services

- CoinDesk - coindesk.com

- CoinTelegraph - cointelegraph.com

- TodayOnChain - todayonchain.com

- NewsBTC – newsbtc.com

- Bitcoin Magazine - bitcoinmagazine.com

- Crypto Slate – cryptoslate.com

- Bitcoin.com – news.bitcoin.com

- Blockonomi – blockonomi.com

Charting Services

◆ TradingView
tradingview.com (best overall, best social)

◆ CoinMarketCap
coinmarketcap.com (simple, easy)

◆ CryptoWatch
cryptowat.ch (very established, best for bots)

◆ CryptoView
cryptoview.com (very customizable)

◆ Coinigy
coinigy.com (great range of pairs and exchanges)

◆ Coin 360
coin360.com (unique UI, check this one out!)

◆ Altrady
altrady.com (scanners, handy tools)

NFT Marketplaces

- OpenSea – opensea.io (general and simple)

- Rarible – rarible.com (general, community-owned)

- Superrare – superrare.com (high end)

- Foundation – foundation.app (for creators and artists)

- Solana – solanart.com (Solana-based NFTs)

YouTube Channels

- Benjamin Cowen
  https://www.youtube.com/channel/UCRvqjQP
  SeaWn-uEx-w0XOIg

- Coin Bureau
  https://www.youtube.com/c/CoinBureau

- Forflies
  https://www.youtube.com/c/Forflies

- DataDash
  https://www.youtube.com/c/DataDash

- Sheldon Evans
  https://www.youtube.com/c/SheldonEvansx

- Lark Davis
  https://www.youtube.com/channel/UCl2oCaw
  8hdR_kbqyqd2klIA

Index

Selected Bibliography

1. Cawrey, D., Lantz, L. (2020). *Mastering Blockchain*. United States: O'Reilly Media.
2. Bulkowski, T. N. (2013). *Fundamental Analysis and Position Trading*. Wiley.
3. "The 7 Best Price Action Patterns Ranked by Reliability." *Samurai Trading Academy*, 3 Nov. 2019, samuraitradingacademy.com/7-best-price-action-
4. Carmen, J. D. (2021, September 27). *Bitcoin and Altcoin market cycles*. The Hub on ACCOINTING.com. Retrieved September 29, 2021, from https://www.accointing.com/the-hub/market-analysis/2021/bitcoin-and-altcoin-market-cycles/.
5. John, A., & Law, J. (2021). *Crypto Technical Analysis* (Vol. 1). Aude Publishing.
6. Varnes, R. S. (2020). *Chart Logic - Technical Analysis Handbook* (Vol. 1). Independently Published.
7. "83 Candlestick Pattern Indicators for TradeStation." *TechnicalTradingIndicators.com*, www.technicaltradingindicators.com/tradestation-indicators/83-candlestick-patterns/.
8. Buntinx, J. P. (2016, December 20). *Top 4 cryptocurrency projects created before bitcoin*. The Merkle News. Retrieved September 29, 2021, from https://themerkle.com/top-4-cryptocurrency-projects-created-ahead-of-bitcoin/.
9. Carmen, J. D. (2021, September 27). *Bitcoin and Altcoin market cycles*. The Hub on ACCOINTING.com. Retrieved September 29, 2021, from https://www.accointing.com/the-hub/market-analysis/2021/bitcoin-and-altcoin-market-cycles/.
10. Raval, S. (2016). *Decentralized Applications: Harnessing Bitcoin's Blockchain Technology*. United States: O'Reilly Media.
11. Algorithmic Trading - Hard Facts." *ClickAlgo*, clickalgo.com/algorithmic-trading - hardfacts#:~:text=Professional%20traders%20do%20not%20have,once%20these%20 conditions%20are%20identified%2C.
12. Laurence, T. (2019). *Blockchain*. For Dummies, a Wiley brand.
13. Berentsen, A., Schar, F. (2020). *Bitcoin, Blockchain, and Cryptoassets: A Comprehensive Introduction*. United States: MIT Press.
14. Alton, Larry. "Why the Fibonacci Sequence Is Your Key to Stock Market Success." *The BiggerPockets Blog | Real Estate Investing & Personal Finance Advice*, 17 Jan. 2018, www.biggerpockets.com/blog/fibonacci-sequence.
15. Infante, R. (2019). *Building Ethereum Dapps: Decentralized Applications on the Ethereum Blockchain*. (n.p.):
16. Manning Dannen, C. (2017). *Introducing Ethereum and Solidity: Foundations of Cryptocurrency and Blockchain Programming for Beginners*. United States: Apress.
17. *Bitcoin and Blockchain: History and Current Applications*. (2020). United Kingdom: CRC Press.

18. Berthou, Adrien. "Token Supply 101: Fundamentals of Token Supply-and Monetary Policy." *Medium*, Medium, 20 Aug. 2020, medium.com/@AdrienBe/token-supply-101-fundamentals-of-token-supply-and-monetary-policy-a20cc761f6ec.

19. "The Best Charting Tools For Crypto Traders." *RSS*, cryptotrader.tax/blog/the-best-charting-tools-for-crypto-traders#:~:text=TradingView%20is%20by%20far%20the,free%20users%20and%20pro%20users.

20. "Cryptocurrency Technical Analysis 101." *OKEx*, www.okex.com/academy/en/cryptocurrency-technical-analysis-101#:~:text=Though%20technical%20analysis%20in%20the,or%20decrease%20in%20the%20future.

21. *Blockchain use cases IN 2021: Real world industry applications*. ConsenSys. (n.d.). Retrieved September 29, 2021, from https://consensys.net/blockchain-use-cases/.

22. *Blockchain use Cases: IBM BLOCKCHAIN*. Blockchain use cases | IBM Blockchain. (2021, July 6). Retrieved September 29, 2021, from https://www.ibm.com/blockchain/use-cases/.

23. Brunton, F. (2020). *Digital Cash: The Unknown History of the Anarchists, Utopians, and Technologists Who Created Cryptocurrency*. United States: Princeton University Press.

24. Ahonen, Elias. *Blockland: 21 Stories of Bitcoin, Blockchain, and Cryptocurrency*. Cryptonumist (Elias Ahonen Inc.), 2021.

25. Voskuil, Eric. *Cryptoeconomics: Fundamental Principles of Bitcoin*.

26. "The 12 Tenets of Sound Fundamental Analysis and Valuation Principles." *Old School Value*, 22 Mar. 2016, www.oldschoolvalue.com/stock-valuation/the-12-tenets-of-sound-fundamental-analysis-and-valuation-principles/.

27. "The 3 Types of White Papers (& When to Use Them)." *Compose.ly*, 29 Sept. 2020, compose.ly/strategy/types-of-white-papers/.

28. *Bitcoin explained - Chapter 7: BITCOINS Scalability*. Investerest. (n.d.). Retrieved September 29, 2021, from https://investerest.vontobel.com/en-dk/articles/13323/bitcoin-explained---chapter-7-bitcoins-scalability/.

29. *Factors affecting cryptocurrency mining profit*. EastShore Mining Devices. (2020, May 6). Retrieved September 29, 2021, from https://www.eastshore.xyz/factors-affecting-cryptocurrency-mining-profit/.

30. Hertig, Alyssa, et al. "Bitcoin Halving 2020, Explained." *CoinDesk*, 17 Dec. 2020, www.coindesk.com/bitcoin-halving-explainer.

31. Chris Douthit, Chris Douthit, and Chris Douthit. "17 Stock Chart Patterns All Traders Should Know." *Option Strategies & Stock Market News*, 3 Apr. 2020, optionstrategiesinsider.com/blog/17-stock-chart-patterns-to-look-for-when-analyzing-stocks/.

32. Binance Academy. (2021, August 18). *What is a blockchain consensus algorithm?* Binance Academy. Retrieved September 29, 2021, from https://academy.binance.com/en/articles/what-is-a-blockchain-consensus-algorithm.

33. Cointelegraph. "What Is A White Paper And How To Write It." *Cointelegraph*, Cointelegraph, 24 Apr. 2018, cointelegraph.com/ico-101/what-is-a-white-paper-and-how-to-write-it.

34. *Banks consume over three times more energy than Bitcoin, according to researcher.* Bitcoinist.com. (2018, August 23). Retrieved September 29, 2021, from https://bitcoinist.com/banks-consume-energy-bitcoin/.

35. "Comprehensive Candlestick Patterns Guide." *Steve Nison's Candlecharts.com*, 5 Oct. 2020, candlecharts.com/candlestick-training/candlestick-patterns/.

36. Conway, Luke. "Bitcoin Halving: What You Need to Know." *Investopedia*, Investopedia, 18 Mar. 2021, www.investopedia.com/bitcoin-halving-4843769#:~:text=A%20Bitcoin%20halving%20event%20is,which%20new%20Bitcoins%20enter%20circulation.

37. "The Deflationary Economics of the Bitcoin Money Supply." *Skalex.io*, 7 May 2020, www.skalex.io/deflationary-economics-bitcoin/.

38. Dorman, Jeff. "Jeff Dorman: Fundamental Investing Is Alive and Well in Crypto." *CoinDesk*, CoinDesk, 5 Feb. 2021, www.coindesk.com/fundamental-investing-alive-well-crypto.

39. Deepti Waghmare. "High-Frequency Trading Comes to Cryptocurrency." *The FinReg Blog*, 2 June 2020, sites.law.duke.edu/thefinregblog/2019/04/24/high-frequency-trading-comes-to-cryptocurrency/.

40. "White Papers Archives." *CoinDesk*, CoinDesk, www.coindesk.com/tag/white-papers.

41. -, E. M., By, -, Edith M.Edith is an investment writer, M., E., & writer, E. is an investment. (2021, February 10). *Edith M.* Forex Academy. Retrieved September 29, 2021, from https://www.forex.academy/detailed-breakdown-of-bitcoins-four-years-cycles/.

42. Bajpai, P. (2021, September 8). *Countries where bitcoin is legal and illegal.* Investopedia. Retrieved September 29, 2021, from https://www.investopedia.com/articles/forex/041515/countries-where-bitcoin-legal-illegal.asp.

43. Cobb, C. (2004). *Cryptography for dummies.* John Wiley & Sons.

44. Edwards, J. (2021, September 21). *Bitcoin's price history.* Investopedia. Retrieved September 29, 2021, from https://www.investopedia.com/articles/forex/121815/bitcoins-price-history.asp.

45. Edwood, F. (2020, October 23). *Why low latency is important for cryptocurrency exchanges, explained.* Cointelegraph. Retrieved September 29, 2021, from https://cointelegraph.com/explained/why-low-latency-is-important-for-cryptocurrency-exchanges-explained.

Endnotes

[i] "Binance Exchange Whitepaper Accessed November 26, 2021. https://www.exodus.com/assets/docs/binance-coin-whitepaper.pdf.

[ii] "Tether: Fiat Currencies on the Bitcoin Blockchain." Accessed November 26, 2021. https://tether.to/wp-content/uploads/2016/06/TetherWhitePaper.pdf.

[iii] "Solana: A New Architecture for a High Performance Blockchain." Accessed November 26, 2021. https://solana.com/solana-whitepaper.pdf.

[iv] "Cardano Ada Whitepapers." Cardano ADA whitepapers. Accessed November 26, 2021. https://whitepaper.io/coin/cardano.

[v] "The Ripple Protocol Consensus Algorithm." Accessed November 26, 2021. https://ripple.com/files/ripple_consensus_whitepaper.pdf.

[vi] "POLKADOT: Vision for a Heterogeneous MULTI-CHAIN Framework." Accessed November 26, 2021. https://polkadot.network/PolkaDotPaper.pdf.

[vii] "Whitepaper - F.HUBSPOTUSERCONTENT30.NET." Accessed November 28, 2021. https://f.hubspotusercontent30.net/hubfs/9304636/PDF/centre-whitepaper.pdf.

"Crypto Research, Data, and Tools." Messari Crypto News. Accessed November 28, 2021. https://messari.io/asset/usd-coin/profile.

"Whitepaper - F.HUBSPOTUSERCONTENT30.NET." Accessed November 28, 2021. https://f.hubspotusercontent30.net/hubfs/9304636/PDF/centre-whitepaper.pdf.

"USD Coin (USDC): Fully Reserved Fiat-Backed Stablecoin." USD Coin (USDC) | Fully Reserved Fiat-Backed Stablecoin. Accessed November 28, 2021. https://www.circle.com/en/usdc.

[viii] "Whitepapers." Ava Labs: Build the Internet of Finance. Accessed November 26, 2021. https://www.avalabs.org/whitepapers.

[ix] Dogecoin. Accessed November 28, 2021. https://dogecoin.com/.

Chohan, Usman W. "A History of Dogecoin." SSRN, December 27, 2017. https://papers.ssrn.com/sol3/papers.cfm?abstract_id=3091219.

[x] "UNISWAP v2 Core." Accessed November 28, 2021. https://uniswap.org/whitepaper.pdf.

"A Short History of Uniswap." Uniswap Protocol, February 11, 2019. https://uniswap.org/blog/uniswap-history.

[xi] "Chainlink." Accessed November 28, 2021. https://research.chain.link/whitepaper-v1.pdf.

[xii] "Home - Polygon | Ethereum's Internet of Blockchains." Accessed November 26, 2021. https://polygon.technology/lightpaper-polygon.pdf.

[xiii] "Behold... the Algorand Whitepaper." Simple Bitcoin Portfolio Tracker. Accessed November 26, 2021. https://coin.fyi/news/algorand/behold-the-algorand-whitepaper-ohysad#!

[xiv] "Elrond Whitepaper." Accessed November 26, 2021. https://elrond.com/assets/files/elrond-whitepaper.pdf.

Elrond. "The Future of Money • Maiar." Maiar. Accessed November 26, 2021. https://maiar.com/.

[xv] "Fintech & Blockchain Projects and Partners." Stellar. Accessed November 26, 2021. https://stellar.org/ecosystem/projects.

Tech in Asia - connecting Asia's startup ecosystem. Accessed November 26, 2021. https://www.techinasia.com/stellar-asia-growth.

"Stellar Consensus Protocol." Stellar. Accessed November 26, 2021. https://www.stellar.org/papers/stellar-consensus-protocol?locale=en.

xvi "Axie Infinity Sales Volume Data, Graphs & Charts / CryptoSlam!" / CryptoSlam! Accessed November 26, 2021.
https://cryptoslam.io/axie-infinity/sales/summary.

Axie Infinity - Axie Infinity. Accessed November 26, 2021. https://whitepaper.axieinfinity.com/.

xvii "White Paper - Decentraland." Accessed November 26, 2021. https://decentraland.org/whitepaper.pdf.

xviii "The Maker Protocol White Paper: Feb 2020." The Maker Protocol White Paper | Feb 2020. Accessed November 26,
2021. https://makerdao.com/en/whitepaper/.

xix "Cosmos Network Ecosystem Overview." DezentralizedFinance.com, July 1, 2021.
https://dezentralizedfinance.com/cosmos-network-ecosystem-overview/.

xx "The Sandbox Whitepaper (August 2020) V2." Accessed November 26, 2021.
https://installers.sandbox.game/The_Sandbox_Whitepaper_2020.pdf.

xxi "Helium." Accessed November 26, 2021. http://whitepaper.helium.com/.

xxii Eosio. "Documentation/Technicalwhitepaper.md at Master · EOSIO/Documentation." GitHub, April 28, 2018.
https://github.com/EOSIO/Documentation/blob/master/TechnicalWhitePaper.md.

xxiii"Enjin Coin." Accessed November 28, 2021. https://cdn.enjin.io/downloads/whitepapers/enjin-coin/en.pdf.

An Aude Publication.

Visit audepublishing.com for free books, discounts, and updates.

per aspera ad astra.

.